CONSTRUCTION ESTIMATING

COMPLETE HANDBOOK

EXCEL ESTIMATING INCLUDED

By Adam Ding

Published by:

 DELMAR
CENGAGE Learning™

www.DeWALT.com/guides

DELMAR
CENGAGE Learning™

**DeWALT Construction Estimating Complete
Handbook: Excel Estimating Included**
Adam Ding

Vice President, Technology and Trade
    Professional Business Unit:
    Gregory L. Clayton

Product Development Manager: Robert Person

Development: Nobina Chakraborti

Director of Marketing: Beth A. Lutz

Director of Building Trades: Taryn McKenzie

Marketing Manager: Marissa Maiella

Production Director: Carolyn Miller

Production Manager: Andrew Crouth

Content Project Manager: Brooke Greenhouse

Art Director: Benjamin Gleeksman

© 2010

For product information and technology assistance, contact us at
Cengage Learning Customer & Sales Support, 1-800-354-9706

For permission to use material from this text or product,
submit all requests online at **www. cengage.com/permissions.**
Further permissions questions can be e-mailed to
permissionrequest@cengage.com

Library of Congress Control Number: 2009937353

ISBN-13: 978-1-4354-9899-0
ISBN-10: 1-4354-9899-2

Delmar
5 Maxwell Drive
Clifton Park, NY 12065-2919
USA

Cengage Learning is a leading provider of customized learning solutions with
office locations around the globe, including Singapore, the United Kingdom,
Australia, Mexico, Brazil, and Japan. Locate your local office at: **international.cengage
.com/region**

Cengage Learning products are represented in Canada by Nelson Education, Ltd.

Visit us at **www.InformationDestination.com**
For more learning solutions, please visit our corporate website at **www.cengage.com**

NOTICE TO THE READER

Printed in Canada
1 2 3 4 5 6 7 12 11 10 09

CONTENTS

iv   CONTENTS

2. QUANTITY TAKEOFF

2. QUANTITY TAKEOFF 29
Quantity Takeoff Procedures. 29
Common Unit Abbreviations 30
Converting Lengths. 30
Calculating Areas 31
Calculating Volumes 32
Unit Conversions. 33
Converting Imperial Measures to Metric Measures. 33
Converting Metric Measures to Imperial Measures. 34
General Quantity Takeoff Worksheet. 35
Estimating Sitework. 36
Estimating Demolition 37
Estimating Earthwork 38
Estimating In-Place, Loose, and Compacted Yards. 39
Calculating Swell and Shrinkage. 40
Swell. 40
Shrinkage and Shrinkage Factor 40
Estimating Example 1 40
Estimating Example 2 40
Estimating Bulk Earthwork—Grid Method 40
Estimating Example 41
Estimating Bulk Earthwork—End Area Method 42
Estimating Example 42
Calculating Excavation Slope 45
Estimating Strip Footing Excavation 46
Estimating Math 46
Estimating Example 46
Estimating Pad Footing Excavation 47
Estimating Math 47
Estimating Example 47
Estimating Foundation Backfill. 47
Estimating Math 47
Estimating Example 1 47
Estimating Example 2 48
Estimating Basement Excavation 48
Estimating Math 48
Estimating Example 49

FOREWORD

WHY THIS BOOK?

How much does it take to build the project? You probably won't know the answer to this question until the job is complete. It's true that the construction industry today faces great challenges: material price volatilities, skilled labor shortage, strict regulations, tight schedule, shoddy drawings. . . . Any one of these can make a seemly lucrative job end up ugly.

Is there anything you can do? Based on real-life experience in estimating hundreds of projects, this book attempts to provide part of the solution. With careful cost estimating efforts, you can at least begin with an accurate budget. Then, even if things go wrong, you could still achieve a decent profit eventually.

Presented in a straight-forward, no-nonsense format, the book covers the following subjects in extensive detail.

- Marketing, bid planning, and scope review
- Quantity takeoff for all trades and divisions
- Pricing of material, labor, and equipment
- Subtrade quote evaluation
- Overhead, profit, and bid procedures
- Cost analysis, value engineering, and change order pricing
- Excel spreadsheet estimating tutorials (a unique feature of this book)

Fully loaded with estimating tips, checklists, worksheets, and data tables, the book walks you through each step of the estimating process and frequently suggests more efficient ways to accomplish the work.

This is not just a book about how to measure quantities. It is aimed to help you take real control over the numbers in your business and make the money you deserve. Yes, you can!

CHAPTER 1

ESTIMATING PREPARATION

How much do you know about your numbers? Estimating is the bloodline of contracting. Without quality estimates, a contractor cannot stay in business. Estimating is where the profits are usually lost, even before the job starts.

Before starting an estimate, you need to have a procedure. In this chapter, the following topics will be covered.

- Overview of the cost estimating process
- Contract relationship and marketing strategy
- Decision to bid and workload planning
- Review of bid documents
- Evaluating document quality
- Measuring building area and preliminary estimating
- Subtrade contacts and bid invitation
- Site investigation worksheet
- Estimating design-build and renovation jobs

Construction Estimating Process

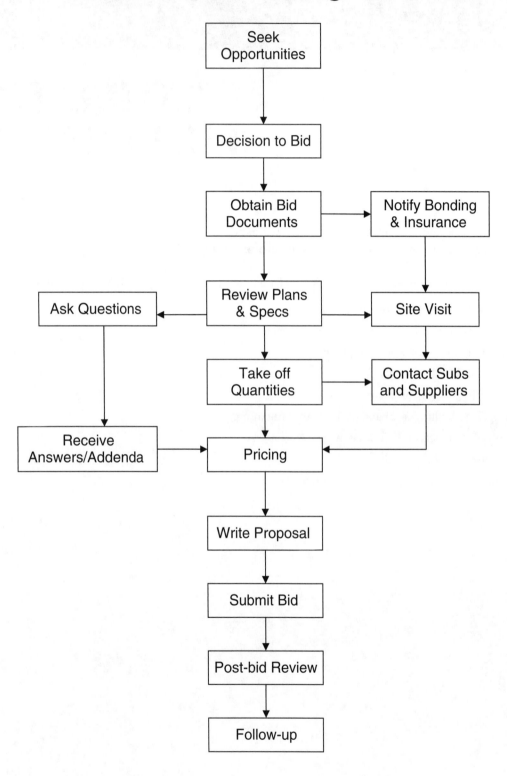

Mechanism of Pricing

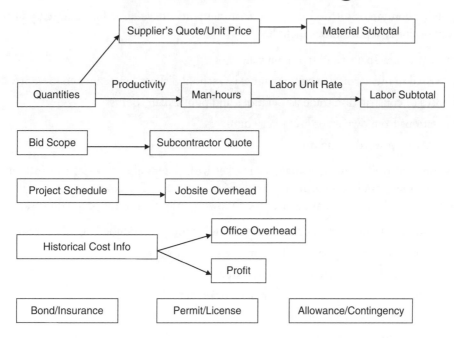

Traditional Contract Relationships

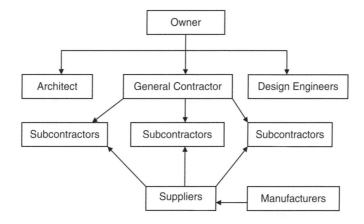

This diagram symbolizes the most common contract relationships in the construction industry. Owners have direct contract relationships with architects, engineers, and general contractors, but not with subcontractors or suppliers. General contractors sign a contract with owners and then award subcontracts.

In recent years, certain variants of this contract relationship have emerged. For example, in a construction management (CM) relationship, there may be no contract relationships between the general contractor and the subtrades. Instead, owners sign contracts directly with subcontractors for each portion of work. The general contractor simply manages the job for the owner, and may or may not perform any of the actual work.

FINDING BUSINESS OPPORTUNITIES

If you are a reputable contractor in the local area, customers may have already requested you to bid projects. Normal sources of business opportunities include:

- Advertisement in newspaper and trade journals
- Bulletins posted in offices of government agencies, school districts, state universities, private colleges
- News services of construction trade associations and public plan rooms
- Direct invitations from owners or architect/engineers
- Business contacts or word of mouth

Often contractors, especially new ones, have to market with potential customers to win their business. You must show them why they should do business with you instead of another contractor. You will have to provide more than a business card. You may want to offer the following information for prequalification.

- Full contact information: company name, address, telephone/fax number, email address
- Company status: corporation, partnership, or sole proprietorship
- Years in business under the current company name
- Names of company officers, principals, partners, or owners
- Type and size of projects you would like to bid
- Type of work you do with your own forces
- Type of work you subcontract to others
- List of jurisdictions where your firm is licensed
- Average annual dollar volume for the past 3 years
- Number of current employees in the office and field
- List of current jobs including contract amounts
- List of completed projects including contract amount, completion date, architect and general contractor names, contacts, and phone numbers
- Three or more business references including company names, contact persons, and phone numbers
- Ability to bond and bonding capacity, bonding agent name, contact person, and phone number
- Insurance agent name, contact person, phone number, and sample certificate of insurance
- Bank name and contact
- Trade associations of which you are a member

By the nature of the job, an estimator is partly a salesperson. Your role is often more than quantity takeoff and pricing. With a progressive attitude, you can make a difference in contributing to the success of your company.

FACTORS TO CONSIDER IN BID DECISION

You will not (and should not) bid every job that comes your way. When deciding whether to bid a job, consider this essential question: Is it a job I am likely to get and make a decent profit from? If the answer is no, then you should decline the invitation to bid. Do not bid work when you lack the skilled workers, the money, or the expertise to handle the job. Challenge yourself with the following hard questions.

- Does the bid due date fit into the current work schedule?
- Is this job similar to what the company normally does (e.g., in terms of type, size, complexity, quality)?
- Do I understand the scope of the work?
- Does the owner have a good reputation? Does the owner have enough money to pay?
- Is the owner using me to do a price check?
- Is this a real job or just a budgeting exercise?
- Are there many contractors bidding the job? Who are they?
- Have I bid against these contractors previously? What was the outcome?
- How far is this job from my office?
- Is the job in my "territory"? Are there licensing issues?
- Am I familiar with the subtrades in that area and local building codes?
- Is the job in a heavy union area? Is my company union or open shop?
- How and when will I expect to receive the payments once the job starts?
- What are the procedures for change orders (if allowed)? How long should I have to wait before getting paid for changes?
- Who is the architect or engineer? Does this person have a solid reputation?
- Are drawings complete? Is the job still in the design phase?
- How many jobs do I already have underway? Among them, how many are near completion? If another job is awarded, will it exceed my bonding capacity? Will staff be available to run the job to finish it on time?
- Do I have to invest in special tools and equipment to do the work?
- Is there anything unusual about the job, such as night shifts or restricted working hours?
- Overall, can I expect a decent profit if I get the job?

Your insurance company will also need to know some project specifics before quoting the insurance and bonds. You should assess the following risk factors.

- Shoring
- Underpinning
- Blasting
- Excavation below 20 feet
- Pile driving
- Wind and flood risks
- Construction complexity

USING A BID PLANNING WORKSHEET

You often bid five or six jobs at the same time, hoping to get at least one. The bidding process requires spending money and time on each bid. To make the efforts worthwhile, you need to set priorities and be organized. Setting priorities means you focus more on the bids that you have a better chance of winning and realizing a profit on. The following worksheets help prioritize your workload planning.

Bid Planning Worksheet

ABC Contracting
1 Main Street
Anytown, USA 00000
(555) 555-1234

Evaluation Factors	Bid #1	Bid #2	Bid #3
Job name			
Bid date			
Size			
Type of construction			
Location			
Similar previous jobs?			
Within performing capacity?			
Number of competitors			
Owner's reputation			
Designer's reputation			
Overall winning chances			

PRE-BID INFORMATION CHECKLIST

ABC Contracting
1 Main Street
Anytown, USA 00000
(555) 555-1234

Job Name: _____

Bid Due Date/Time: _____ # of Copies: _____

Bid Delivery Format: ❏ Fax   ❏ E-mail   ❏ Personal   ❏ Courier

Send Bid To: ❏ Architect   ❏ Owner, Contact Info: _____

Job Location: _____ Tax Rate: _____

Construction Type: _____   ❏ New Construction   ❏ Renovation

Building Area: _____ Perimeter: _____ # of Floors: _____ Height: _____

Site Acres: _____ Paved Area: _____ Landscaped Area: _____

Functional Units: _____ Quality of Finishing: _____ Approximate Value: _____

Owner's Name: _____ Architect's Name: _____

Pre-bid Meeting: Date/Time: _____ Location: _____ Participants: _____

Names of Competitions: _____

Project Start Date: _____ Project Finish Date: _____

Liquidated Damages: _____

Labor Conditions: ❏ Prevailing Wages   ❏ Union   ❏ Open Shop

Permit Costs/Development Fees: ❏ Include in Proposal   ❏ Exclude (by Owner)

Bond: ❏ Bid Bonds   ❏ Payment & Performance Bonds

Insurance: ❏ Builder's Risk   ❏ Liability Insurance   ❏ Owner's Wrap-up Insurance

Material Testing ❏ Include in Bid Proposal   ❏ Exclude (by Owner)

Work Done by Owner's Contractor (Exclude from the Bid): _____

Cost Breakdown: _____ Cash Allowance: _____ Alternates: _____

Download this form at www.**DeWALT**.com/guides

REVIEWING BID DOCUMENTS

In order to get everything right from the very beginning, you need to effectively review the bidding documents. The following steps are helpful in your review.

1. *Read instructions to bidders*: Read every word in this important document to know what is required to submit bid proposals.

2. *Check documents package*: Count the drawings to ensure you have everything you need. Do the same with the spec book.

3. *Examine drawings*: Thoroughly go through one drawing sheet after another, including all detail sections. Circle unfamiliar details with red pencil. Try to get the overall picture of the job.

4. *Read specs*: Read all sections at least once. Pay attention to agreement and supplementary/special conditions for items affecting the costs, such as working hours and parking. In trade sections, look for the types of materials or systems specified and whether substitutions are allowed. If certain sections are not applicable to the project at hand, then ask the architect for clarifications.

5. *Take notes*: While reviewing drawings and specs, create a list of items affecting the costs. Group them by trades, not by drawing sheets.

Try to go beyond blueprints and visualize what it takes to construct the building. Not all information will be shown on drawings, and some of these items may involve heavy equipment or intensive labor.

Document Package Checklist

ABC Contracting
1 Main Street
Anytown, USA 00000
(555) 555-1234

Job Name: _____   Checked By: _____

❏ Civil Drawings
- Survey
- Demolition
- Paving and Drainage
- Utilities
- Site Electrical
- Soil Report
- Landscaping & Irrigation

❏ Architectural Drawings
- Floor Plan
- Exterior/Interior Elevation
- Roof Plan
- Sections & Details

- Interior Design

❏ Structural Drawings
- Foundation
- Floor Framing
- Roof Framing
- Sections & Details

❏ Mechanical & Electrical
- Fire Protection
- HVAC
- Plumbing
- Electrical

❏ Specifications Book

EVALUATING QUALITY OF BID DOCUMENTS

Sometimes during the bidding process, certain documents may be in the design stage and not ready for pricing. To avoid last-minute surprises that your subtrades are not bidding due to lack of information, use the following checklist to determine if all plans are ready for your pricing instead of heading back to the drawing board.

ABC Contracting
1 Main Street
Anytown, USA 00000
(555) 555-1234

Site Drawings Should Indicate:
- ❒ Overall dimensions and north arrow
- ❒ Adjoining streets and property lines
- ❒ Benchmarks and reference points
- ❒ Locations of existing and proposed work
- ❒ Existing and proposed contours
- ❒ Final grade elevations at all building corners and along building perimeters
- ❒ Walks, drives, or other surface features including slopes and cross slopes of accessible routes
- ❒ Locations and critical elevations of utilities, wells, disposal fields
- ❒ Parking facilities and storm drainage system
- ❒ Sanitary sewer, water supply and distribution, fire underground system; locations of fire hydrants (existing and new)
- ❒ Coordination of utilities with the mechanical, electrical, and plumbing plans
- ❒ Flow lines and invert information on manholes
- ❒ Soil investigation data

Architectural Drawings Should Indicate:
Floor Plan
- ❒ Appropriate scale (1/8 inch per foot or greater)
- ❒ "Key plan" to designate the project portion to which each sheet applies
- ❒ Dimensions, room titles, orientation north arrow, floor elevations
- ❒ Coordination with structural, mechanical, and electrical plans
- ❒ Supplementary drawings for congested areas, toilet rooms, equipment rooms, etc.

Roof Plan
- ❒ Same scale as floor plan
- ❒ Materials and system used
- ❒ Locations of valleys, ridges, cants, saddles, crickets, gutters, down spouts
- ❒ Projections through roof such as skylights, chimneys, exhaust ducts or vents, penthouses
- ❒ Changes in roof elevation, direction, and amount of slopes

(continues)

Download this form at **www.DeWALT.com/guides**

Elevation Drawings

- ❏ Same scale as floor plan
- ❏ Sectionalized to correspond to floor plan
- ❏ Complete exterior of the building
- ❏ Grade elevations, vertical dimension to floors, ceilings, and roof

Section and Detail Drawings

- ❏ Appropriate scale for clarity
- ❏ Dimensions and levels from roof parapet to top of footings
- ❏ Typical floor, wall, ceiling, and roof construction (assemblies)
- ❏ Door and window schedule
- ❏ Details of door/window heads, jambs, sills, mullions, transoms
- ❏ Room finish schedule
- ❏ Interior room elevations
- ❏ Millwork schedule and details
- ❏ Stair risers, treads, landings, rails, and finishes
- ❏ Typical and special exterior and interior trims
- ❏ List of equipment and fixtures
- ❏ Special construction conditions or architectural features

Structural Drawings Should Indicate:
Foundation Plans

- ❏ Same scale as architectural floor plan
- ❏ Schedule or details showing the size, shape, material, reinforcing, depths and elevations of footings and piers
- ❏ Schedules and details for columns, beams, walls, slabs, and openings
- ❏ All the dimensions necessary for building layout

Framing Plans

- ❏ Same scale as architectural floor plan
- ❏ Elevations and locations for columns, beams, girders, joists, trusses, studs, bolts, anchors, bracings, slabs, and reinforcing details
- ❏ Details for lintels, purlins, trusses, bridging, etc.
- ❏ Sufficient details, schedules, and notes for all structural elements
- ❏ Table of design loads

Mechanical and Electrical Drawings Should Indicate:
Plumbing Plans

- ❏ Scale no smaller than architectural plans
- ❏ Temporary piping
- ❏ Foundation drain line hookups
- ❏ Storm and sewer lines

(continues)

❏ Complete water distribution system

❏ Locations of all plumbing fixtures and equipment

❏ Sewage disposal system

❏ Sewage and vent lines

❏ Gas supply and distribution

❏ Necessary details, isometric diagrams, schedules, and legends for all fixtures, equipment, pipe and fitting types, sizes, and materials

❏ Plumbing riser diagrams and plans

❏ Vents through the roof and coordination with roof plan

❏ Roof drainage and overflow system

❏ If there is no site utility plan, include a plumbing site plan to show location, type, size, and extent of exterior lines, connections, and equipment

❏ Demolition of existing system and connection to new system

Fire Protection Plans

❏ Same scale as architectural plans

❏ Water supply to the building with water flow test information

❏ Outside control valves and fire department connection

❏ Sprinkler and standpipe risers, fire hose cabinets, and building zone control valve locations

❏ Occupancy/hazard class of each area or room, sprinkler design density, and design area of sprinkler coverage

❏ Sprinkler heads, type, nominal orifice size, and temperature rating

❏ Activation appliances and emergency abort button

❏ Interconnections with other building system (fire alarm, fan shutdown, etc.)

❏ All partitions and fire-rated walls

❏ Location and size of concealed spaces, closets, attics, and bathrooms

❏ Enclosures in which no sprinklers are installed

❏ Soffits or known ceiling obstructions

❏ Sprinkler piping, mains, size, and location for coordination

❏ Control valves and check valves, flow switches, water gongs, gauges, etc.

❏ Sprinkler test stations including provisions for test water removal

❏ Fire pumps (if required) including size and flow requirements

❏ Design calculation remote area

❏ Demolition of existing system and connection to new system

HVAC Plans

❏ Same scale as architectural plans

❏ Separate plumbing plans and electrical plans

(*continues*)

- ❏ Partitions and room layouts, fire- and smoke-rated partitions
- ❏ Rated capacity, efficiency, and operating conditions for all equipment
- ❏ Ductwork and piping layout including size, types, pressure class
- ❏ Necessary details, sections, schedules, and notes to show the work scope
- ❏ Building heating and cooling loads (in BTUs/hr), temperature differentials used, and rated capacity of heating units
- ❏ Location of fire and smoke dampers, grilles, outlets, etc.
- ❏ Coil and tube pull areas
- ❏ Required code clearance areas
- ❏ Balancing dampers, splitter dampers, volume extractors, balancing valves, thermometers, pressure gauges, instrument-flow fittings, and instrument-access panels required for balancing
- ❏ Plumbing and electrical connections, system controls
- ❏ Demolition of existing system and connection to new system

Electrical Drawings

- ❏ Same scale as architectural plans, printed full size
- ❏ Names and uses of all rooms, north arrow, and door swings
- ❏ Consist of lighting, power, and auxiliary systems
- ❏ Single line diagram
- ❏ Branch circuit diagram
- ❏ Panel schedules
- ❏ Light fixture and similar electrical equipment schedules with power data
- ❏ Legends and details
- ❏ All connections, permanent or temporary, inside and outside
- ❏ Locations and sizes of all conduits, cables, and wiring
- ❏ Names and capacities of special outlets
- ❏ Location and details of switchboards, motor control centers, power panels, lighting panels, and other equipment
- ❏ Locations of fire alarm appliances and control panels
- ❏ Locations, connections, and controls of signals, speakers, clocks, telephones, fire alarms, and other special systems
- ❏ Required code clearance areas of electrical equipment
- ❏ Interlocks with other systems (fire sprinkler, HVAC, etc.)
- ❏ Demolition of existing system and connection to new system

Specifications Should Indicate:

- ❏ Front-end documents (bid invitation and instructions, contract, general conditions, supplementary conditions, bid forms, others)
- ❏ Project-specific information rather than generic specs

(continues)

❏   Full description of approved materials and systems

❏   Manufacturers' names, products brands, or catalog numbers

❏   Required performance criteria for all materials and assemblies

❏   Installation procedures, cleanup methods, and inspections

❏   Equipment supplied and/or installed by others

Final Considerations:

Are these documents complete enough for pricing? ____Yes ____ No

Are drawings and specifications coordinated? ____Yes ____ No

Are documents in compliance with governing building codes? ____Yes ____ No

Issues to be resolved with architect or engineer:

_____

_____

_____

_____

_____

_____

_____

_____

_____

_____

_____

_____

_____

_____

_____

_____

CREATING A QUESTION LIST FOR DOCUMENT REVIEW

If time is a vital issue, then you must first work from a "big picture" perspective. What information can you extract from the drawings and specs in 20 minutes? Try to answer the following preliminary questions.

Site Drawings and Specs

- Is it an existing site or a new site?
- How much earthwork needs to be done?
- Is there any offsite roadwork?
- Is the building to be serviced with complete utilities, for example, water and sewer?
- What are the soil conditions according to the geo-tech report?
- Can the excavated material be reused for backfill?
- Is there any site electrical and lighting?
- Are there any site-specific features, such as fountains or gazebos?

Architectural Drawings and Specs

- Is there an existing building to be demolished, renovated, or expanded?
- What shape is the new building footprint? How many square feet total?
- Is the architect's number for building area correct? Is the architect measuring from the inside corner of the building, and counting balconies in or out?
- How many floors are planned, and what is on each floor? What is the square footage for each floor?
- What is the distance between floors, and the ceiling height?
- Where are mechanical and electrical rooms on each floor?
- What is the building perimeter on each floor?
- Where is the building entrance?
- What is the name of each room, and how big is each room?
- How high is the exterior wall, and what is it made of?
- What are the principal exterior finish materials?
- What types of doors and windows are planned, and is there a schedule?
- What types of roof systems are specified?
- Are there any balconies, canopies, and walkways?
- What is the partition wall made of, and is it load bearing?
- What is the floor finish, and is there a finish schedule?
- Are there any elevators, stairs, rails, fireplaces, chimneys?
- What types of fittings, fixtures, or equipment must be furnished and installed? What is the scope for the owner-hired contractor?

Structural Drawings and Specs

- What types of structural systems are used?
- How deep is the foundation and is there any special drainage system?
- Is the foundation wall cast-in-place concrete or masonry? Are there any changes in wall height due to varying site elevations?

- What is the concrete strength required?
- How high is the roof?
- Is there any floor below grade, and how deep is it?
- Are the floors made of concrete, steel, or wood?
- What is the roof structure made of?
- How is the structure fire protected? Any spray fireproofing?
- Will any heavy equipment such as cranes be required to do the work?
- What type of construction is planned for each floor?
- Is the lowest floor concrete slab or wood framing with a crawl space?
- Are upper floors wood framing, open web steel joist/decking, concrete suspended slab, or hollow-core precast concrete?
- How will the structural details affect mechanical and electrical trades (e.g., sprinkler heads, ductwork, pipes, conduits, and fixtures)?

Mechanical and Electrical Drawings and Specs

- What types of mechanical/electrical systems are used?
- What types of materials are specified for equipment, piping, ductwork, wiring, or fixtures?
- How will the mechanical and electrical systems affect the site construction?
- How will the mechanical and electrical systems affect the building construction?
- What type of electrical service does the mechanical equipment require?
- Is the mechanical contractor to provide wiring for controls, motors, disconnect switches, motor starters, etc.?

Specifications

- Instruction to bidders
- Owner, architect, engineer
- Chain of command and responsibility
- Bonding and insurance requirements
- Supplementary and special conditions
- Completion dates and liquidated damages
- Payment terms, schedule, and holdback
- Change order procedures
- Dispute resolution
- Cash allowances
- Alternates
- Maintenance, turnover, and training
- Names of manufacturers or subcontractors for building systems (e.g., specified curtain wall and storefront suppliers, millwork contractors, approved roofers and painters)

COMPLETING A PROJECT SCOPE QUICK CHECKLIST

Following is a 5-minute basic checklist for job scope. It is not intended to replace the detailed review of contract documents.

Project Scope Quick Checklist

ABC Contracting
1 Main Street
Anytown, USA 00000
(555) 555-1234

Lower Floor: _____ Concrete_____ Basement _____ Crawl space

Upper Floors: _____ Wood framed _____ Steel deck _____ Suspended concrete _____ Hollow core

Roof Structure: _____ Wood framing/truss _____ Steel deck _____ Precast concrete

Exterior Walls: _____ Siding _____ Stone _____ Block _____ Brick _____ Stucco _____ Concrete

_____ Trims

Exterior Openings: _____ Curtain wall _____ Storefront _____ Windows _____ Doors

Roof Finish: _____ Shingle _____ Built up _____ Single ply _____ Metal

Building Projection: _____ Soffit _____ Balcony _____ Patio _____ Parapet _____ Canopy

_____ Sidewalk

Interior Walls: _____ Drywall _____ Masonry _____ Concrete _____ Glazed _____ Doors

Floor Finish: _____ Carpet _____ VCT _____ Sheet vinyl _____ Tile _____ Laminate _____ Hardwood

Finished Ceiling: _____ Drywall _____ Acoustical T-bar _____ Wood

Exposed Ceiling: _____ Steel _____ Concrete _____ Wood _____ Painted_____ Primed

_____ Spray insulation

Wall Finish: _____ Paint _____ Tile _____ Vinyl _____ MDF _____ Trims _____ Moldings

Specialties: _____ Washroom accessories _____ Cabinets _____ Vanity _____ Appliance _____ Fireplace

_____ Closet shelving _____ Shower door _____ Blinds _____ Others

Conveying system: _____ Stairways _____ Elevator _____ Lifts _____ Chutes

M/E: _____ Plumbing _____ HVAC _____ Sprinkler _____ Electrical

Site development: _____ Earthwork_____ Paving _____ Utilities _____ Landscaping/irrigation

---

Download this form at **www.DeWALT.com/guides**

MAKING DETAILED DOCUMENT REVIEW NOTES

If time is not of the essence, then make a thorough review of the drawings and specs. The best way to do so is to make detailed notes, as shown in the following example.

Job Name: All American School
Notes Made By: ABC CONTRACTING
Date: January 1, 2009

Trades	Reference	Review Notes
Sitework	C-1	Certified Survey & As Builts
		Material Testing to be Included
	C-2	Demo Existing Building
	C-3	Temp Access Road Required
	C-4	Offsite Work & Retention Pond
		MOT & Traffic Signalization
	RFI	Soil Report? Existing Survey?
	C-5	Site Fire Line & DDCV
		Need to Send Plans to ABC Sitework
Landscape/Irrigation	RFI	Owner's Allowance? Need Clarify
	C-2	Existing Tree Relocation
	L-1	Plant Count to Compare w/ Schedule
	I-1	Jack & Bore for Irrigation Conduits
Concrete	C-2	Site Dumpster & Screen Wall Footing
	C-3	Site Sidewalk, Need QTO
	A-101	Building Sidewalk
		Define Brick Paver Scope
	S-201	Concrete Beams & Columns
	P-2	Concrete for Plumbing Trenches
	E-101	Concrete for Electrical Conduits
Masonry	C-2	Dumpster Enclosure and Site Screen Wall
	A-104	Exterior Brick Veneer, Contact Supplier
		Detailed Masonry QTO
Metals	C-7	Dumpster Gate & Bollards
	A-104	Stainless Steel Handrail, Quote
	A-201	Roof Ladder

(continues)

Trades	Reference	Review Notes
		Preengineered Trusses, Metal/Wood?
	A-301	Support Framing for RTU
	S-100	Signed & Sealed Shop Drawings
Wood & Plastics	A-100	Fire-rated Plywood for Exterior Canopy
	A-104	Exterior Wood Decking
	A-106	Millwork Furnished by Owner, incl. Installation
	A-107	Fypon Trim, Vinyl Soffit
Division 7	A-105	Built-up Roofing w/ Hatch, Tile Roofing
	Specs	List of Approved Roofers
		Stainless Steel Sheet Metal
		20-Year Warranty Labor & Materials
	A-201	Aluminum Panel on Parapet Wall
	RFI	Need Clarification on Fireproofing
Doors	A-202	Hollow Metal Door, Wood Door
		Overhead Door, Automatic Door
Glass	A-202	Glass for Automatic Doors
	Specs	List of Approved Curtain Wall System
	A-601	Interior Glass and Mirrors on Wall
	A-104	Exterior Windows
	S-100	Signed & Sealed Shop Drawings
	RFI	Impact-resistant Glass?
Drywall/Stucco	A-201	Trusses
	S-100	Signed & Sealed Shop Drawings
	A-601	Interior Demising Wall to Reach Roof Deck
	A-104	Stucco w/EIFS Trims
Ceiling/Flooring	A-603	Need Flooring QTO by Each Type
Paint	RFI	Is Interior Wall to Be Painted?
	A-501	Vinyl Wall Covering

(continues)

Trades	Reference	Review Notes
Division 10-14	A-303	Fire Extinguishers
	C-2	Site Chain Link Fence
	A-401	Metal Awnings
	A-601	Ceiling Access Panels
	A-101	Mailboxes, Trash Receptacles, Bike Racks
HVAC	M-1	Sheet Metal Ductwork for Base Bid
	M-2	Condensate Drainage Piping
	RFI	Owner Furnishing RTU?
Fire	C-5	Site Fire Line
	RFI	Fire Sprinkler Drawings
		Fire Protection for Exterior Canopy?
Plumbing	C-5	Two Grease Traps
	P-1	Water Heater
	A-105	Roof Drainage Connection
Electrical	C-2	Power for Monument Signs
	C-5	Power for Lift Station
	I-1	Power for Irrigation
	E-100	Fire Alarm System
	RFI	Panels and Fixtures to be Supplied by Owner?
	Specs	Temp Power
	SE-100	Concrete Bases for Lighting Poles

MEASURING BUILDING GROSS FLOOR AREA

Architects usually include the gross floor area (GFA) on the drawings, but their calculation could be different from yours (e.g., they may only measure from the inside corners of the building). Use the following steps to measure the GFA for construction purposes.

1. Go around the building perimeter, and measure areas from the exterior wall corners.
2. Find out how many individual floors the job has (basement, main floor, upper floors, loft, penthouse, etc.). Remember to include the upper stories.
3. Subdivide each floor into smaller segments that are easier to measure. Break down the areas into shapes, such as rectangles, squares, triangles, circles, and semicircles.

4. Calculate the area of each shape separately and total them to get the area for each floor.

5. Total the area of each floor to get the total building gross area.

Technology such as digitizers or electronic on-screen takeoff has made the area measurement much easier and more visual. Regardless if you use computers, the following guidelines for measuring areas can be helpful.

- Do not always scale drawings, because the drawings you see could have been reduced in size. Always try to use the dimensions as shown.

- Do not add room areas to figure the total building area. Some schematics show bedrooms as 15" × 12", bathrooms as 9' × 6' and so forth, but wall thicknesses are not included in these dimensions.

- Measure non-A/C or nonheated areas separately (e.g., garage, driveway, covered walkways, canopies, patios, balconies, terrace, soffits). Depending on your company rules, these areas may or may not be part of the overall GFA (e.g., some builders will take half of the balcony decks to be included in their area measurements).

- It might be useful to know the separate numbers for residential areas versus commercial areas, below grade areas versus above grade areas, condo unit interior areas versus common function areas, wood framed areas versus concrete structure areas, to name a few.

The following is a sample area takeoff.

Floor	Area	Perimeter	Floor Height	Ceiling Height
	sq ft	ft	ft	ft
Parking	1,064	169	11.0	9.0
Level 1	9,515	527	11.5	8.0
Level 2	9,256	426	9.0	8.0
M/E Level	1,626	184	9.3	8.0
Total Area	**21,461 sq ft**			

PREPARING A PRELIMINARY ESTIMATE

It is always important to prepare a detailed and accurate estimate, but before you dive into the details of the project, it is helpful to have a general feel of what the ballpark figure will likely be. A preliminary estimate can be prepared based on years of estimating experience and adequate historical job cost data. There are two ways to determine this figure.

1. Apply unit prices to the number of functional units (e.g., the number of residential condos, hotel rooms, school students, hospital beds). For example, the total construction cost for a high-rise apartment tower of 60 luxury units is as follows:

$$60 \times \$450,000 = \$27,000,000$$

2. Apply unit prices to different functional areas (e.g., areas for parking, residential living, commercial retail or common facilities). For example, in the same apartment building:

Parking area: 20,000 sq ft × $50 = $1,000,000
Living area: 130,000 sq ft × $200 = $26,000,000
Total costs: $1,000,000 + $26,000,000 = $27,000,000

You may consider using both methods for cross-checking. The resulting number is the starting point of your estimate, but should not be the ending point. This preliminary number will be verified later using accurate detailed estimating.

SAMPLE RFI TO ARCHITECT

Request for Information

Project Name: All American School

Date: January 1, 2009

TO: XYZ Architects, Tel: (999) 123-4567, Fax: (999) 123-4568

FROM: Estimator, ABC Contracting, Tel: (111) 123-4567, Fax: (111) 123-4568

1) Please provide soil report if available.

2) Please clarify if we need to include landscape and irrigation.

3) Please clarify if the upper floor deck is to be sprayed with fireproofing.

4) Please confirm the glass is to be impact resistant.

5) Please confirm the interior wall is to be painted.

6) Please provide fire sprinkler drawings if available.

7) Please confirm if the owner is supplying roof-top HVAC units, electrical panels, and lighting fixtures.

Please call if you have any questions.

CONTACTING SUBCONTRACTORS AND SUPPLIERS

To bid a job, most likely you will need subcontractors or suppliers. In some cases, you are only allowed to use the ones specified by owners or designers. Otherwise, your local trade associations could recommend a list of their members by specialties.

One reliable method is to contact material suppliers for the individuals they trust. In other words, contact pre-cast storm structure suppliers to find out site subs, ready-mix suppliers to find out concrete subs, block, brick, or rebar suppliers to find out masonry subs, joist and decking suppliers to find out steel subs, HVAC unit suppliers to find out mechanical subs, and lighting fixture or switchgear suppliers to find out electrical subs.

The names of some subs may not be familiar to you. For verification purposes, you can "screen" them by asking for a list of professional references about the jobs they have done before, and check out each reference; or ask them to provide supporting documents such as insurance certificates and business licenses.

EXAMPLE LIST OF COMMON SUBTRADES

No one is truly a "jack of all trades." The decision of whether to use subcontractors on a job depends on several factors.

- Are you licensed to engage in that specialty (e.g., M/E)?
- Who is cheaper to do the work, the sub or yourself?
- Do you have enough resources and expertise to self-perform the work?
- What are quality requirements (or finishing class) for the work?

Depending on the type of job (e.g., residential or commercial), the arrangement of subcontracting might be different. Consider the following examples when determining use of subcontractors.

- Sitework sub (could be the combination of a few subtrades, including earthwork, utility, paving, curbs, line painting)
- Landscaping and irrigation subs (could be two separate companies)
- Concrete or formwork contractor (normally residential subs will not furnish the concrete or reinforcing material)
- Concrete mix supplier
- Rebar supplier (sometimes this supplier also installs)
- Masonry sub
- Structural steel sub (mostly for commercial or institutional jobs)
- Miscellaneous metal sub
- Railing and deck sub
- Framing material and trims supplier
- Residential framer (only providing the labor)
- Rough or finish carpenter (e.g., some subs may specialize in door and accessories installation)
- Millwork sub (especially for custom commercial millwork)
- Residential casework supplier (mostly standard kitchen and bath cabinets)
- Roof or siding sub (normally one trade)
- Insulation sub (drywall subs might include their own insulation)
- Hollow metal/wood doors and toilet accessories supplier
- Special door supplier (for doors other than hollow metal or wood)
- Vinyl windows supplier (material only)
- Glass sub (installed, including curtain wall, storefront, aluminum windows)
- Drywall sub (could also be doing stucco, etc.)
- EIFS/stucco sub
- Acoustical ceiling sub
- Flooring sub
- Painting sub
- Residential appliance supplier

- Swimming pool sub
- Mechanical sub (may include plumbing, HVAC, and fire sprinkler)
- Electrical sub

For more information, please verify with your specific job requirements.

WRITING A BID INVITATION

A bid invitation, the official document containing general bid information, should be created to communicate with your subs and suppliers. A well-written invitation saves hours over the phone in explaining the job information to subs.

Review drawings and specs before writing the invitation so you know what trades will be involved for the scope of work. You can send bid invitations by fax, mail, or email, but only those relevant trades should be notified. Note the following example.

Invitation to Bid

ABC Contracting
1 Main Street
Anytown, USA 00000
(555) 555-1234

From: _____

To: _____

Project Name: _____

Project Address: _____

Pricing Due Date & Time: _____

Scope of Work: _____

Plans and Specs Are Available at: _____

Cost Breakdown: _____

Alternates: _____

Bid Contact: _____

Your Feedback: _____   ❒ Bid   ❒ No Bid

Download this form at **www.DeWALT.com/guides**

DISTRIBUTING BID DOCUMENTS

No one can bid a job without having reviewed the documents. The reality is, however, that some subs or suppliers do not want to pay for drawings to bid jobs. There are a few ways to solve this problem.

- Use a public plan room or construction association service. Give them a set of documents so that their member companies can have access through there.

- If drawings are in electronic format, upload them online. Through a secured web server (with user name and password), subcontractors can check out documents on the internet.

- Store the documents with a reprographics company, and then inform your subs that plans are available from there. They can go there and purchase whatever they want.

- Set aside a work space in your company where subs can view plans.

Whatever you do, please make sure subs will have access to the whole set of documents, instead of just portions of plans. Otherwise, they might turn in incorrect or incomplete bids and claim that they lack information. For example, if you only provide a concrete sub with the architectural and structural sheets, he or she may exclude the concrete work associated with plumbing trenches and electrical conduits.

If subcontractors insist you provide the documents, then have them fill out the following sheet.

Job Name: _____

We need the following documents in order to bid the above project:

Drawings	Sheets/Sections
Civil	
Landscape/Irrigation	
Architectural	
Structural	
Interior Design	
Plumbing	
Fire Protection	
HVAC	
Electrical	
Specifications	

Please deliver the above documents to:

Contact Person: _____

Company Name: _____

Delivery Address: _____

Telephone Number: _____

Charge to Account: _____

USING A SITE VISIT WORKSHEET

Before you bid a job, you always want to visit the site. Site conditions might be different from what you see on the drawings. Get familiar with bidding documents before you go. Your findings become an important part of the bid documentation. During your visit, complete the following worksheet.

Site Visit Worksheet

ABC Contracting
1 Main Street
Anytown, USA 00000
(555) 555-1234

Date: _____

By: _____ Accompany: _____

Job Name: _____ Address: _____

Directions: _____ Road Conditions: _____

Site Access: _____

Site Description: _____

Soil Conditions: _____ Water Table & Drainage: _____

Hazardous Material: _____

Availability of Utilities (water, sewer, electricity): _____

Future Locations for Dumpsters/toilets/trailers: _____

Material Storage: _____ Parking: _____

If Existing Building, Describe Situation: _____

Neighborhood Information: _____

Local Building Codes: _____

Situation with Subs: _____ Labor Conditions: _____

Availability of Housing: _____ Equipment to Be Used: _____

Additional Comments: _____

Sketches/photos Attached: _____

Download this form at **www.DeWALT.com/guides**

ESTIMATING DESIGN-BUILD JOBS

Sometimes you are required to estimate a job based on nothing more than a few sketches. These projects, with almost no drawings or specs available, are often called "design-build" jobs (opposed to "plans-specs" jobs where complete design exists). The contractor is required to team up with a designer to price the job.

Taking the following steps can be helpful

1. Thoroughly study the available documents.
2. Communicate with the owners to find out what they want (e.g., the structural systems and the class of finishes they prefer).
3. Attend the pre-bid meeting and visit the site.
4. Make sure everyone involved (architect, engineer, subtrades, suppliers, bonding, insurance) knows this is a design-build project.
5. Perform material takeoff and pricing based on how the work should be done, not limited to what the drawings show.
6. Allow enough money to cover the gaps between subtrade quotes, that is, what subtrades will not cover, such as mechanical/electrical excavation.
7. Ask your subtrades to detail in their proposals what assumptions they made (e.g., material grade, installation method) in pricing the work.
8. Take subtrade's proposal clarifications, visualize other problematic issues, and list the qualifications for your proposal to the owner.
9. Include all soft costs (e.g., design fees including architectural and engineering, reimbursable expenses, permit, development charges).

ESTIMATING RENOVATION JOBS

Renovation jobs involve demolition, addition, or alteration done to an existing structure. They have more uncertain factors than jobs that are purely new construction. A site visit is always mandatory to examine the current condition of the building as well as locating existing utilities.

Especially check the following issues.

- Are old drawings or specs (including as-built) for existing building available? Which contractor built the existing structure?
- Where are existing utilities located? Is it difficult to gain access?
- Is the building poorly or well maintained? Are there any hazardous materials (e.g., asbestos) present?
- Are there any unusual job conditions (e.g., high ceilings, flooded basement, crawl space, occupant use)?
- Is it difficult to move tools, equipment, and material around?
- Are existing fixtures, piping, and equipment to be removed or relocated?
- Are cutting and patching of existing surfaces part of the contract?
- Are there any dust control and noise abatement requirements?

- Are there any working-hour restrictions? Is the existing facility to remain in operation during the construction period?
- Will temporary shoring be required (e.g., to support the existing roof)?
- Have you increased labor and material rates to cover the unknowns?

CHAPTER 2

QUANTITY TAKEOFF

Accurate quantity takeoff is an essential factor when figuring an estimate. Although the methods of takeoff vary among companies, it is almost always important to be consistent and organized.

In this chapter, the following topics will be covered.

- Principles of quantity takeoff and useful worksheets
- Length, area, and volume calculations
- Quantity takeoff methods for all trade divisions (from earthwork to mechanical and electrical)
- Solved estimating examples with illustrative sketches
- Comprehensive quantity takeoff checklists for each trade

QUANTITY TAKEOFF PROCEDURES

The term *takeoff* means you take the information off the documents and translate it into a list of items with quantities. It can be done in three steps.

1. *Define takeoff scope:* What needs to be taken off? Thoroughly study plans and specs to find out the answer. For unclear details, ask the architect/owner rather than making wild assumptions.
2. *Measure each item:* Use dimensions as specified, and do not scale drawings unless it is necessary. Mark the drawings for the items you took off, because you may not finish the work without being interrupted.
3. *Record quantities:* Make detailed reference as to which sheet you found the items, and where they exist in the building. Record your quantities with drawing number, detail number, and grid reference. It is also important to keep different items separate.

Quantity takeoffs are required for self-performed work; but why take off quantities for subtrades? Essentially, it is a good "yard-stick". In doing so, you familiarize yourself with the scope of the project. When quotes come in, you can determine whether they are reasonable. For some material suppliers, they will need your quantities before quoting the job.

COMMON UNIT ABBREVIATIONS

Unit abbreviations help to clarify the meanings of takeoff quantities and also to save time. The following is a list of commonly used abbreviations.

bf = board foot

cf = cubic foot

cy = cubic yard

ea = each

ft = feet

lb = pound

lf = linear foot

l/s = lump sum

sf = square foot

sfca = square foot of contact area

sy = square yard

sq = square (100 sf)

When dealing with subtrades, it is important to work with the same type of unit of measure. For example, flooring contractors may casually refer to the area as "500 yards," when they mean "500 square yards." Similarly, earthwork contractors will simply use "yards" when referring to "cubic yards." Sometimes, the situation is not very clear. For example, an awning contractor may give you a unit price of "$250 per foot." You must determine whether this is per square foot or per linear foot.

CONVERTING LENGTHS

For easier estimating, convert all lengths to linear feet in decimal format. Simply divide inches by 12, for example, 8 inches/12 = 0.67 ft. Sometimes you need to convert fractions of an inch into feet as well, such as 1/2 inch or 5/8 inch. Remember that 1/8 inch = 0.01 ft. For example, 3/8 inch = 3 × 0.01 = 0.03 ft, thus, 3 ft 8-3/8 inches is 3 ft + 0.67 ft + 0.03 ft = 3.70 ft. The following conversion chart is provided for your convenience.

Inches	Inches in Decimals	Feet in Decimals
1/8	0.125	0.01
1/4	0.250	0.02
3/8	0.375	0.03
1/2	0.500	0.04
5/8	0.625	0.05
3/4	0.750	0.06
7/8	0.875	0.07
1	1.000	0.08
2	2.000	0.17
3	3.000	0.25

Inches	Inches in Decimals	Feet in Decimals
4	4.000	0.33
5	5.000	0.42
6	6.000	0.50
7	7.000	0.58
8	8.000	0.67
9	9.000	0.75
10	10.000	0.83
11	11.000	0.92

CALCULATING AREAS

The following table shows how to calculate areas for common shapes.

Shape	Formula
	Rectangle, square, or parallelogram Area = a × b
	Triangle Area = 1/2 × a × b
	Circle Area = 3.1416 × a^2 = 0.7854 × b^2
	Trapezoid Area = c × 1/2 × (a + b)

For example, to calculate the area of a circle with 2 ft diameter, from the table provided, $0.7854 \times 2^2 = 3.14$ sf (note the diameter is 2 ft and the radius is 1 ft).

CALCULATING VOLUMES

Most volume calculations are simply length × width × height, though sometimes it can get a bit more complicated. Refer to the following table for formulas.

Shape	Formula
	Cylinder Volume = 3.1416 × a/2 × a/2 × b = 0.7854 × a² × b
	Pyramid Volume = 1/3 × a × b × c
	Cone Volume = 1/3 × 3.1416 × b × b × c = 1.0472 × b² × c or Volume = 0.3518 × a² × c

For example, the excavation volume for a round light pole base (5 ft deep, 2 ft in diameter) is $0.7854 \times 2^2 \times 5 = 15.7$ cf.

UNIT CONVERSIONS

Following are a few of the most commonly used unit conversions.

Length: 1 foot = 12 inches

1 yard = 3 feet

Area: 1 square foot = 144 square inches

1 square yard = 9 square feet

1 square = 100 square feet

1 acre = 43,560 square feet

Volume: 1 cubic yard = 27 cubic feet

Weight: 1 ton (imperial) = 2,000 pounds

For example, the concrete needed to pour a 4 in slab, 70 ft long and 50 ft wide, can be figured as:

$$\text{Volume} = \text{Length} \times \text{Width} \times \text{Thickness} = 70 \text{ ft} \times 50 \text{ ft} \times 4 \text{ in}/12$$
$$= 1,167 \text{ cf or } 1,167/27 = 43.2 \text{ cy (without waste)}$$

When working in countries like Canada, where metric systems are used, it is important to know the conversions between imperial and metric units (e.g., feet and meters, pounds and kilograms, U.S. gallons and liters).

Converting Imperial Measures to Metric Measures

Length		
From	**To**	**Multiply By**
inch (in)	meter (m)	0.0254
inch (in)	millimeter (mm)	25.4
foot (ft)	meter (m)	0.3048
yard (yd)	meter (m)	0.9144

Area		
From	**To**	**Multiply By**
square foot (sq ft)	square meter (sq m)	0.0929
square yard (sq yd)	square meter (sq m)	0.8361
square (sq)	square meter (sq m)	9.2903
acre (ac)	square meter (sq m)	4047
acre (ac)	hectare (ha)	0.4047

Volume		
From	**To**	**Multiply By**
cubic foot (cu ft)	cubic meter (cu m)	0.0283
cubic foot (cu ft)	liter (l)	28.317

From	To	Multiply By
cubic yard (cu yd)	cubic meter (cu m)	0.7646
American gallon (gal)	liter (l)	3.7853

Weight/Density		
From	To	Multiply By
pound (lb)	kilogram (kg)	0.4536
short ton, 2,000 lb	kilogram (kg)	907.1848
pound per linear foot (lb/ft)	kilogram per meter (kg/m)	1.488
pound per square inch (psi)	kilopascal (kPa)	6.894
pound per square inch (psi)	megapascal (MPa)	0.0069
pound per cubic foot (pcf)	kilogram per cubic meter (kg/m^3)	16.02

Converting Metric Measures to Imperial Measures

Length		
From	To	Multiply By
millimeter (mm)	foot (ft)	0.0033
millimeter (mm)	inch (in)	0.0394
meter (m)	inch (in)	39.37
meter (m)	foot (ft)	3.2808
meter (m)	yard (yd)	1.0936

Area		
From	To	Multiply By
square meter (sq m)	square foot (sq ft)	10.7639
square meter (sq m)	square yard (sq yd)	1.196
hectare (ha)	acre (ac)	2.471
hectare (ha)	square foot (sq ft)	107639

Volume		
From	To	Multiply By
cubic meter (cu m)	cubic foot (cu ft)	35.3145
cubic meter (cu m)	cubic yard (cu yd)	1.308
cubic meter (cu m)	board foot (bf)	423.783
cubic meter (cu m)	American gallon (gal)	264.2
liter (l)	American gallon (gal)	0.2642

Weight/Density		
From	**To**	**Multiply By**
kilogram (kg)	pound (lb)	2.2046
metric ton	short ton, 2,000 lb	1.1023
kilogram per meter (kg/m)	pound per linear foot (lb/ft)	0.672
kilopascal (kPa)	pound per square inch (psi)	0.145
megapascal (MPa)	pound per square inch (psi)	145
kilogram per cubic meter (kg/m^3)	pound per cubic foot (pcf)	0.0624

GENERAL QUANTITY TAKEOFF WORKSHEET

Project: _____   Date: _____

Location: _____   Page _____ of _____

Trade: _____   Takeoff by _____

Item Description	Details				Extension
	Pieces	**Length**	**Width**	**Height**	
Total					

ESTIMATING SITEWORK

Sitework can be broken down into the following major work packages. (Subcontractors may perform one, several, or even all of these.)

- Demolition
- Earthwork
- Utilities (storm, water, sanitary, gas and fire underground)
- Paving and curbs
- Line painting and exterior signage
- Landscaping (hard and soft)
- Irrigation

Certain items are typically not covered by the site subcontractor and thus will be excluded from the quote. These include the following:

- Earthwork for the building itself (e.g., excavation for foundation, excavation for mechanical and electrical trades, slab prep)
- Building interior demolition
- Site concrete work (sidewalk, walkways, retaining walls)
- Site electrical and lighting
- Site features (e.g., gazebos, arbors, fountains)

You can ask other subtrades to cover these costs, or estimate them yourself. Even if you have to use some "plug" numbers, it is better than leaving them out completely.

ESTIMATING DEMOLITION

Demolition can range from removing a couple of windows to dismantling a complete building. The earthwork contractor may include site demolition in the quote, but typically not interior selective demolition. The following is a checklist to takeoff demolition (with measurement units).

Estimating Demolition Checklist

- ❐ Demolish existing building (sf or l/s)
- ❐ Remove trees (ea)
- ❐ Remove fence (lf)
- ❐ Saw cut (lf)
- ❐ Remove asphalt paving (sy or cy)
- ❐ Remove curb (lf or cy)
- ❐ Remove concrete slabs and sidewalk (sf)
- ❐ Demolish concrete foundation, columns, beams, and staircases (cy)
- ❐ Demolish floors, walls, ceilings, and roofs with finishes (sf)
- ❐ Demolish structural steel columns, beams, and joists (ea)
- ❐ Demolish doors, windows, millwork, specialty items (ea)
- ❐ Cutting and patching (l/s)
- ❐ Temporary fencing (lf)
- ❐ Temporary partitions (sf)
- ❐ Shoring and engineering (l/s)
- ❐ Hazardous material removal (l/s)
- ❐ Dumping (l/s, cy, or tons)

Download this form at **www.DEWALT.com/guides**

ESTIMATING EARTHWORK

Use the following checklist to takeoff earthwork (with measurement units).

Estimating Earthwork Checklist

- ❐ Clear and grub (acres)
- ❐ Dewatering (l/s)
- ❐ Topsoil removal (cy)
- ❐ Excavation (cy)
- ❐ Rough grading (sy)
- ❐ Shoring and underpinning (l/s)
- ❐ Backfill (cy)
- ❐ Import fill (cy)
- ❐ Place and compact (cy)
- ❐ Soil stabilization (l/s)
- ❐ Building foundation excavation (cy)
- ❐ Mechanical and electrical excavation (cy)
- ❐ Building slab prep (cy)
- ❐ Haul away and disposal (cy)
- ❐ Soil testing (l/s)
- ❐ Jacking, boring, and piling (lf)
- ❐ Support and protection (l/s)
- ❐ Silt fence or turbidity barrier (lf)

Download this form at **www.DEWALT.com/guides**

Estimating In-Place, Loose, and Compacted Yards

To estimate earthwork, you will need to know the following three terms.

- *In-place yards:* The original volume of natural soils before disturbance (e.g., 1 cy); it is also called bank yards.

- *Loose yards:* The increased volume of loose soils after digging (e.g., 1.2 cy dirt from the original 1 cy of earth).

- *Compacted yards:* The decreased volume of soil after it is backfilled and compacted (e.g., a resulting 0.85 cy after compacting the 1.2 cy of loose soil).

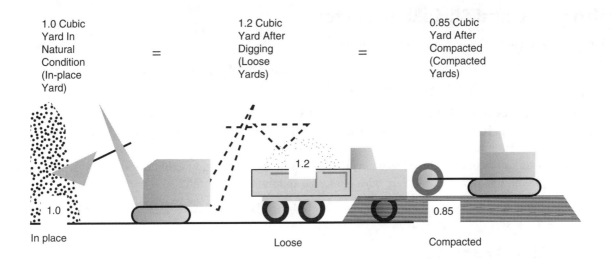

CALCULATING SWELL AND SHRINKAGE

Swell

Swell is the soil volume increase resulting from excavation.

$$\text{Swell (\%)} = (\text{Loose Yards/In-place Yards} - 1) \times 100$$
$$\text{Load Factor} = \text{In-place Yards/Loose Yards}$$

In the previous example, swell is $(1.2 \text{ yards}/1.0 \text{ yard} - 1) \times 100\% = 20\%$, whereas the load factor is $1.0 \text{ yard}/1.2 \text{ yards} = 0.83$, or 83%. Please note: Load Factor $= 1/(1 + \text{Swell})$.

Shrinkage and Shrinkage Factor

Shrinkage is the soil volume decrease from fill compaction.

$$\text{Shrinkage (\%)} = (1 - \text{Compacted Yards/In-place Yards}) \times 100$$
$$\text{Shrinkage Factor} = 1 - \text{Shrinkage}$$

In the previous example, shrinkage is $(1 - 0.85 \text{ yard}/1.0 \text{ yard}) \times 100\% = 15\%$, whereas the shrinkage factor is $1 - 15\% = 85\%$. It is important to understand that shrinkage refers to the in-place yards, not loose yards directly available for fill.

In estimating earthwork, you need to determine the swell and shrinkage factors (if they are not known) and account for them. Actual factors will vary with grain size and moisture content.

$$\text{Excavation Formula: Loose Yards} = \text{In-Place Yards} \times (1 + \text{Swell}).$$

Estimating Example 1

Your calculations show you need to excavate 500 cy of earth for building foundation, which is in-place yards. Suppose soil swell is 20% (or load factor is 0.83), then actual loose yards from excavation is $500 \times (1 + 20\%) = 600$ cy.

Compaction Formula Because Compaction Yards = In-Place Yards $\times$ (1 − Shrinkage) and Loose Yards = In-Place Yards $\times$ (1 + Swell), loose yards required for backfill = Compaction Yards/(1 − Shrinkage) $\times$ (1 + Swell).

Estimating Example 2

Your calculations show you need to backfill 450 cy of earth for building foundation, which is compaction yards. If soil shrinkage is 15% (or shrinkage factor is 0.85), it means the original in-place yards would have been $450/(1 - 15\%) = 530$ cy. In other words, to backfill 450 cy, you will need to use loose yards of $530 \times (1 + 20\%) = 636$ cy. Even if the original excavated 600 cy are all suitable for backfill, you still need to import $636 - 600 = 36$ cy.

ESTIMATING BULK EARTHWORK—GRID METHOD

Normally earthwork estimators today will use a specialized computer software package to take off massive cut and fill volumes, for the sake of accuracy and speed. They digitize the site area, trace existing and proposed contour lines, and specify spot elevations for building and paving. Then the computer software will do the calculations automatically and draw three-dimensional graphs to simulate how the work will be done in the field.

If you decide to manually calculate the volumes, there are at least two methods—the grid method and the end area method. We discuss the grid method first, and follow with the end area method.

The grid method is easier to use when figuring parking lots and site leveling types of projects. You simply divide the earthwork areas into small grids. The sizes of these grids can vary from 10 sf to 50 sf. If the terrain varies greatly in elevations, then using small grids will ensure accuracy.

Take the following steps to calculate earthwork using the grid method.

1.   Visually study the site drawing to determine if it is a cut or fill.

2.   Pick the grid sizes and draw them on the plan.

3.   Use old and new contour lines to determine existing and proposed elevations at each corner of each grid.

4.   Calculate the volumes of cut or fill required for each grid.

5.   Add the grids together for total cut and fill volumes. The result can be an import fill or export haul away.

Estimating Example

In the drawing provided, grid sizes are 50 ft × 50 ft. Contour lines and grid elevations shown are for existing elevations. The entire site must be leveled to an elevation of 100 ft.

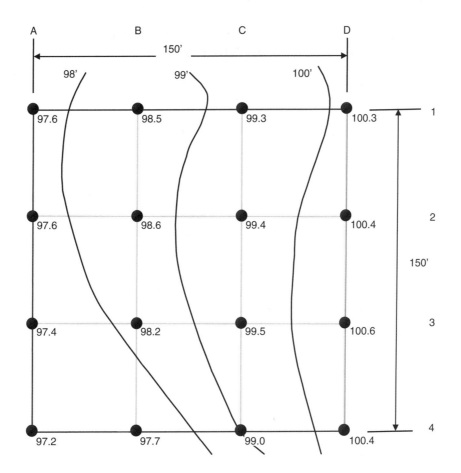

The site seems to be a fill. There are nine grids total. Pick the first one (A1-B1-B2-A2) to start. The following shows the graphic representation of this grid.

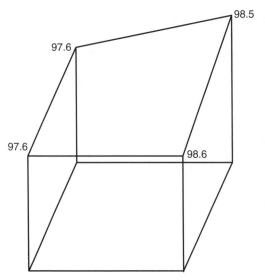

The average existing elevation is (97.6 + 97.6 + 98.5 + 98.6)/4 = 98.08 ft.
Proposed elevation is 100 ft, so the fill height is 100 − 98.08 = 1.92 ft.
The fill volume for this grid is 50 ft × 50 ft × 1.92 ft = 4,800 cf (that is, 178 cy).
You can do the same with the eight remaining grids. The results are as follows:

178	97	14
190	100	2
220	130	12

The total fill volume is approximately 942 cy (net quantity).

ESTIMATING BULK EARTHWORK—END AREA METHOD

The end area method is primarily used in long and narrow tract sites such as roadwork. The site is divided into equally spaced stations. The calculation is a bit more complicated than the grid method, but equally effective.

Take the following steps to calculate earthwork using the end area method.

1. Break down the site into stations at regular intervals (e.g., 100 ft intervals).
2. Take cross sections at each station. Draw a profile based on elevation change.
3. Calculate the cut and fill area for each cross section.
4. Obtain the volume of earthwork between sections by taking the average of the end areas at each station (in square feet) multiplied by the distance between sections (in feet). Convert the result to cubic yards.

Estimating Example

In the drawing provided, four stations are spaced 100 ft apart. Solid contour lines are for existing and dotted contour lines are for proposed elevations.

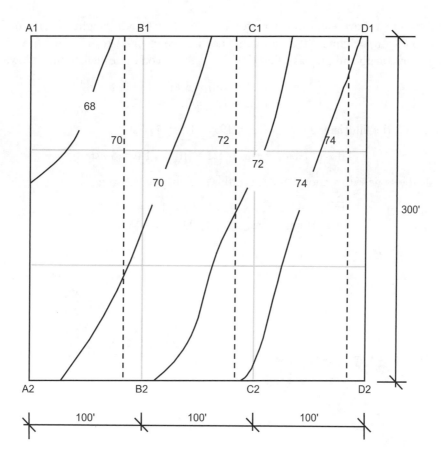

Take station A1-A2, for example. The existing elevation for point A2 is 69.5 ft and point A1 is 66.7 ft. The proposed elevations for both points are 68.5 ft. Draw the profile of the section as shown. Connect the points for existing elevations to form two triangular areas.

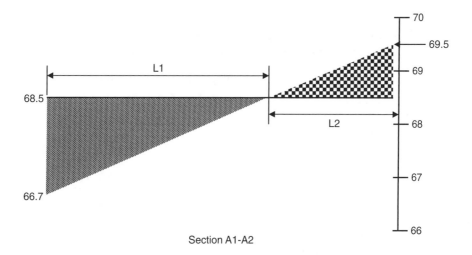

Section A1-A2

The left triangle where it shows elevation increases from 66.7 ft to 68.5 ft (height 1.8 ft) is for fill and the other (drop from 69.5 ft to 68.5 ft, height 1.0 ft) is for cut. The distance from A1 to A2 is 300 ft, which is the total of L1 and L2. Using simple math and the basis that L1:L2 = 1.8:1.0, you can calculate the following:

$$L1 = 1.8/(1.8 + 1.0) \times 300 = 193 \text{ ft}$$
$$L2 = 1.0/(1.8 + 1.0) \times 300 = 107 \text{ ft}$$

So for station A1-A2, the fill triangle area is $1/2 \times 193 \times 1.8 = 174$ sf.

The cut triangle area is $1/2 \times 107 \times 1.0 = 54$ sf.

Now you can do the same with station B1-B2. For this station, you will get:

Fill triangle area is $1/2 \times 210 \times 1.8 = 189$ sf.
Cut triangle area is $1/2 \times 90 \times 1.5 = 68$ sf.

Section B1- B2

To calculate earthwork volume between station A1/A2 and B1/B2,

Fill volume = $(174 + 189)/2 \times 100 = 18,150$ cf (that is, 672 cy).
Cut volume = $(54 + 68)/2 \times 100 = 6,100$ cf (that is, 226 cy).

If you do the same thing for two other sections, you will get the following volumes.

Sections	Fill	Cut
A1/A2-B1/B2	672	226
B1/B2-C1/C2	567	441
C1/C2-D1/D2	215	791
Total	**1,454**	**1,458**

The result seems to be a balanced site with minimal haul away or import fill. It is important to remember, however, that fill volume calculated is compacted yards and cut volume calculated is in-place yards. Therefore, some importing of soils is required.

CALCULATING EXCAVATION SLOPE

Note the following important facts.

- Excavation slopes are calculated by dividing *horizontal* distances by *vertical* distances. This definition is different from the "slopes" in other trades.

- Data above only apply to excavations less than 20 ft deep.

- Very few soils are stable enough for type A or better. When no information is available, it is safe to assume the soil is type C.

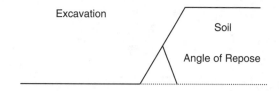

Types of Soil	Maximum Slope (H:V)	Angle of Repose
Stable Rock	Vertical	90°
Type A (Hard and solid soil)	3/4:1	53°
Type B (Soil likely to crack or crumble)	1:1	45°
Type C (Soft, sandy, filled or loose soil)	1-1/2:1	34°

ESTIMATING STRIP FOOTING EXCAVATION

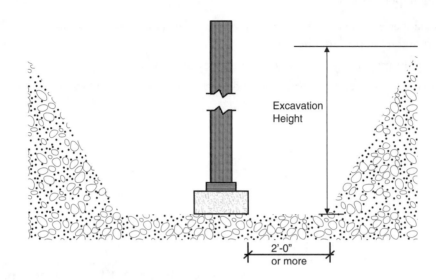

Estimating Math

To account for workspace and the sloped excavation volume:

Excavation Volume (net) = (Footing Width + 2 × Workspace Width) × Excavation Height × Footing Length
+ Excavation Height × Excavation Slope × Excavation Height × Footing Length

Sometimes when the footing is shallow, excavation slope may be not required.

Estimating Example

Footing is 30 ft long × 2 ft wide; work space is 2 ft and excavation slope is 1:1 (45 degree angle). Excavation height is 7 ft. Then excavation volume (net) is (2 + 2 × 2) × 7 × 30 + 7 × 1 × 7 × 30 = 2,730 cf (that is, 2,730/27 = 101.2 cy).

ESTIMATING PAD FOOTING EXCAVATION

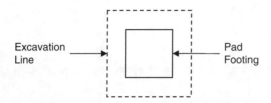

Estimating Math

If excavation is shallow and no slope is required, then do the following:

Excavation Volume (net) = (Pad Width + 2 × Workspace) × (Pad Length + 2 × Workspace)
                          × Excavation Height

If excavation is deep, then you need to account for both work space and slopes:

Excavation Volume (net) = (Pad Width + 2 × Workspace + Excavation Height × Excavation Slope)
                          × (Pad Length + 2 × Workspace + Excavation Height × Excavation Slope)
                          × Excavation Height

Estimating Example

The pad is 4 ft long, 4 ft wide, and 12 inches high; work space is 2 ft, excavation height is 1.5 ft, and no slope is required. Then excavation volume (net) is (4 + 2 × 2) × (4 + 2 × 2) × 1.5 = 96 cf (that is, 96/27 = 3.6 cy).

ESTIMATING FOUNDATION BACKFILL

Estimating Math

For strip footing with foundation wall:

Backfill Volume (net) = Excavation Volume − Footing Concrete − Wall Concrete

For isolated pad footing:

Backfill Volume (net) = Excavation Volume − Pad Concrete

Estimating Example 1

Footing is 30 ft long, 2 ft wide, and 12 inches high. Foundation wall is also 30 ft long, 10 inches thick, and 8 ft high. Work space is 2 ft and excavation slope is 1:1 (45 degree angle). Excavation height is 7 ft.

Excavation volume is (see previous examples on how to estimate excavation) (2 + 2 × 2) × 7 × 30 + 7 × 1 × 7 × 30 = 2,730 cf.

Footing concrete volume is 30 × 2 × 1 = 60 cf.
Wall concrete volume is 30 × 0.83 × 8 = 200 cf.
Backfill volume (net) is 2,730 − 60 − 200 = 2,470 cf.

This volume, however, is compacted yards. To figure the actual backfill required, convert to loose yards. Normally you can use 40% more fill material to cover shrinkage, waste, and so forth, so the actual backfill volume is 2,470 × (1 + 40%) = 3,458 cf, or 128 cy.

Estimating Example 2

The pad is 4 ft long, 4 ft wide, and 12 inches high; work space is 2 ft, excavation height is 1.5 ft, and no slope is required.

Excavation volume is (see previous example on how to estimate excavation) (4 + 2 × 2) × (4 + 2 × 2) × 1.5 = 96 cf.

Concrete volume is 4 × 4 × 1 = 16 cf.
Backfill volume is 96 − 16 = 80 cf (that is, 80/27 = 3.0 cy).

This volume is also compacted yards. To figure the actual backfill required, convert to loose yards. Normally you can use 40% more fill material, so the actual backfill volume is 3.0 × (1 + 40%) = 4.2 cy.

Keep in mind that the excavated soil may not be reused for backfill. For example, the engineer may want crushed stone rather than dirt to ensure water drainage, so native soil would not be reused in this situation.

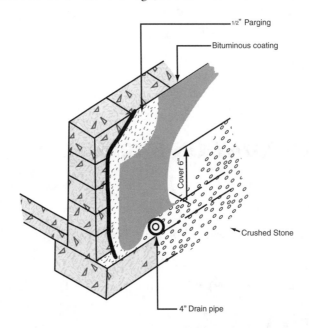

ESTIMATING BASEMENT EXCAVATION
Estimating Math

Excavation Volume (net) = (Building Area + Building Perimeter × Excavation Slope × Excavation Depth) × Excavation Depth
Backfill Volume (net) = Excavation Volume − Building Area × Excavation Depth

When measuring the building perimeter, make sure the distances include footing width projection and work space. Also, the excavation is performed from the grade level to the bottom of foundation (not just from basement floor to ceiling).

Estimating Example

Consider a 35 ft $\times$ 60 ft building with basement excavation average depth of 9 ft. Excavation slope is given as 1:2 (or angle of repose at 60 degrees) for very stable soil.

Building area: $35 \times 60 = 2{,}100$ sf

Building perimeter: $(35 + 60) \times 2 = 190$ ft

Excavation depth: 9 ft

Basement excavation volume (net):

$(2{,}100 + 190 \times 1/2 \times 9) \times 9 = 26{,}595$ cf, or $26{,}595/27 = 985$ cy

Backfill volume (net): $26{,}595 - 2{,}100 \times 9 = 7{,}695$ cf, or $7{,}695/27 = 285$ cy

Actual backfill loose yard material: $285 \times (1 + 40\%) = 399$ cy

Estimating Mechanical/Electrical Excavation

Underground mechanical pipes and electrical conduits will require excavation of the earth, burial of the pipes/conduits, and sand backfill. Occasionally it may require concrete encasement. Documents should be thoroughly studied to verify what is needed, because the work is potentially extensive.

Estimating Example

An underground waterline is 150 ft. Trench is 2 ft wide and 4 ft deep.

> Excavation (net): $150 \times 2 \times 4 = 1{,}200$ cf (that is, $1{,}200/27 = 44.5$ cy).
> Backfill (net): $150 \times 2 \times 4 = 1{,}200$ cf (that is, $1{,}200/27 = 44.5$ cy).
> Actual backfill loose yard material is $44.5 \times (1 + 40\%) = 62.3$ cy.

Please note this calculation does not consider slope volumes for excavation, which could be significant for deeper trenches (e.g., 10 ft or more).

ESTIMATING SITE UTILITIES

Use the following checklist for utilities (with measurement units).

Estimating Site Utilities Checklist

- ❐ Mobilization and demobilization (l/s)
- ❐ Surveying and as-builts (l/s)

Storm Drainage

- ❐ Excavation and backfill (cy)
- ❐ Manhole demolition (ea)
- ❐ Pipe demolition (lf)
- ❐ Manholes (ea), listing different types
- ❐ Catch basins (ea)
- ❐ Culverts (lf)
- ❐ Pipes (lf)—note different materials and dimensions (e.g., 18 or 30 inch RCP or PVC)
- ❐ Tie-ins (ea)
- ❐ Roof drain connection (ea)
- ❐ Foundation drain connection (ea)
- ❐ Sump pumps (ea)

Sanitary Sewer

- ❐ Excavation and backfill (cy)
- ❐ Open cut and repair (sy)
- ❐ Manholes (ea), listing different types
- ❐ Cleanouts (ea)
- ❐ Pipes (lf)—note different materials and dimensions (e.g., 4 inch versus 6 inch, DIP versus PVC)
- ❐ Septic tanks (ea)
- ❐ Grease traps (ea)
- ❐ Lift stations (ea)

Water

- ❐ Excavation and backfill (cy)
- ❐ Open cut and repair (sy)
- ❐ Pipes (lf)—note different materials and dimensions (e.g., 2 inch versus 3 ich, copper versus PVC)
- ❐ Fittings (ea), including tees, valves, etc.
- ❐ Backflow preventer (ea)
- ❐ Testing and balance (l/s)

Fire Underground

- ❐ Excavation and backfill (cy)
- ❐ Open cut and repair (sy)
- ❐ Pipes (lf)—note different materials and dimensions (e.g., 4 inch versus 6 inch, DIP versus PVC)
- ❐ Fittings (ea), including tees, valves, etc.
- ❐ Fire department connection (FDC) (ea)
- ❐ Fire hydrants (ea)

Site electrical conduits and lighting are done by electricians and thus will not be part of the site servicing package.

Drawings often show the work by utility companies (e.g., sewer, water, gas, electrical connections). It is important to check with the owner to confirm whether utility company charges are part of the contract, as these costs can be significant and you probably will not obtain price quotes until the job actually starts.

ESTIMATING PAVING

The paving portion of site contractor's quote normally includes:

Estimating Paving Checklist
❏   Asphalt paving (sy), including subbase, gravel, and asphalt
❏   Curbs (lf), listing each type, including curb base prep
❏   Parking bumpers (ea)
❏   Concrete driveway apron, handicapped ramp, or sidewalk (sf)
❏   Site signage (ea) for traffic and parking control
❏   Line painting (ea or lf)

Download this form at **www.DEWALT.com/guides**

Although paving is normally done at the end of project, it is important to keep the schedule in mind, as winter seasons might be too cold to carry out the work.

Estimating Example

A 200 ft × 30 ft parking lot is paved with 1.5 inch asphalt, 6 inch base crushed gravel, and 8 inch subbase pit run.

> Asphalt paving area is 200 × 30 = 6,000 sf, or 6,000/9 = 667 sy.
> Base crushed gravel = 6,000 × 0.5 = 3,000 cf, or 3,000/27 = 111.2 cy.
> Subbase pit run = 6,000 × 0.67 = 4,000 cf, or 4,000/27 = 148.2 cy.

It is common to allow as much as 40% extra for base and subbase fill material.

> So actual base crush gravel needed is 111.2 × (1 + 40%) = 156 cy.
> Actual subbase pit run is 148.2 × (1 + 40%) = 208 cy.

ESTIMATING LANDSCAPING AND IRRIGATION

Use the following checklist to takeoff landscaping (with measurement units).

Estimating Landscaping and Irrigation Checklist

Soft Landscaping

- ❏ Trees (ea)
- ❏ Shrubs (ea)
- ❏ Sod (sf)
- ❏ Seed (sf)
- ❏ Mulch (sf)
- ❏ Top soil (cy)
- ❏ Fertilizing (sf)
- ❏ Edgings (lf)
- ❏ Maintenance and warranty (l/s)

Irrigation

- ❏ Sleeves (lf)
- ❏ Pipes (lf)
- ❏ Fittings (ea)
- ❏ Sprinkler heads (ea)
- ❏ Planter drain (lf)

Download this form at **www.DEWALT.com/guides**

Hard landscaping items such as pavers, planter walls, and fountains should be counted and evaluated individually. Separate quotes may be necessary.

ESTIMATING CONCRETE WORK

Depending on the job scope, concrete estimates may include the following items.

Estimating Concrete Work Checklist

- ❐ Continuous footing (footers) or grade beams
- ❐ Pad footings (pads) or pile caps
- ❐ Foundation walls and retaining walls
- ❐ Slabs on grade and curbs
- ❐ Suspended slab and slab bands
- ❐ Topping on metal or wood decks
- ❐ Tie beams and columns
- ❐ Stairs (suspended or on ground)
- ❐ Miscellaneous concrete (e.g., fill to metal stairs, bollards, column bases)

Essentially, concrete estimate is the takeoff of the following (with material, labor, and equipment):

- ❐ Formwork (sfca)
- ❐ Concrete (cy)
- ❐ Reinforcing (lb or ton)
- ❐ Curing and finishing (sf)
- ❐ Miscellaneous items (e.g., layout, fine grading, insulation, damp proofing, soil treatment, slab vapor barrier, caulking of saw-cut joints)

Foundation excavation and backfill, slab base, may already be completed by an earthwork subtrader. If not, include these items when estimating concrete work. You may find the following advice helpful.

- ❐ Check all plans for complete information, not architectural and structural sheets alone. For example, equipment pads and conduit encasement may only show on mechanical and electrical drawings.
- ❐ Always read concrete specs and structural notes for information such as aggregate sizes, strengths, admixtures, and finishing.
- ❐ Carefully plan the method to place concrete: direct chute, bucket, crane, pumping and conveyor belts, for example. Visualize how many pours are required: foundation, slab on grade, walls, columns, suspended slabs, slab bands. When there are several floors, plan it floor by floor and consider coordination with other trades.
- ❐ For concrete volumes, do not deduct small opening areas (e.g., doors, windows, penetrations) for floors and walls. Allow additional formwork and finishing (chamfer strips, etc.) for such openings.
- ❐ For every concrete item exposed (e.g., column corners, exterior walls, stair walls), allow enough finishing costs (sandblast, rubbing, patching).
- ❐ Go through your estimate for possible omissions (e.g., the rebar unloading, field testing, heating for pouring in winter, water reducers or accelerators, setting embeds, grouting and caulking for tilt-up wall panels).

ESTIMATING FORMWORK

For the sake of simplicity, formwork is usually taken off by square feet of contact area (SFCA). Be sure to account for the following miscellaneous formwork items as well (either separately or as part of formwork unit price).

- Formwork hardware
- Keyway and inserts
- Form clamps and bracing
- Scaffolding and shoring
- Expansion joints
- Construction joints or bulkheads
- Form release agent
- Form removal and cleaning
- Chamfer strip

If you decide to use board feet as the measurement unit for estimating formwork, consider the following rules of thumb.

Formwork	Factor
Strip Footing	2.0 bf/sfca
Pad Footing	2.5 bf/sfca
Columns	3.0 bf/sfca
Foundation Wall	2.5 bf/sfca
Wall over 12 ft High	3.5 bf/sfca
Stairs on Grade	2.5 bf/sfca
Stairs Suspended	6.0 bf/sfca

ESTIMATING CONCRETE STRIP FOOTING

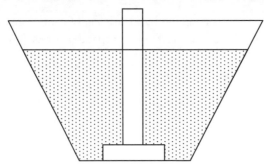

The takeoff of strip footings starts with measuring their lengths and includes:

- Footing layout (l/s)
- Excavation (cy)
- Fine grading (sf)
- Formwork (sfca)
- Concrete (cy)
- Rebar (lb or ton)
- Embeds (ea)
- Backfill (cy)

Estimating Math for Formwork and Concrete

$$\text{Formwork} = (\text{Footing Length} + \text{Footing Width}) \times \text{Footing Height} \times 2$$
$$\text{Concrete} = \text{Footing Length} \times \text{Footing Width} \times \text{Footing Height}$$

Estimating Example

Footing is 30 ft long, 2 ft wide, and 12 inches high.

$$\text{Formwork} = (30 + 2) \times 1 \times 2 = 64 \text{ sfca}$$
$$\text{Concrete} = 30 \times 2 \times 1 = 60 \text{ cf (that is, } 60/27 = 2.3 \text{ cy)}$$

Allow 10% waste, concrete volume is $2.3 \times (1 + 10\%) = 2.6$ cy.

ESTIMATING CONCRETE WALL

The wall takeoff starts with measuring wall lengths and heights to reach a wall area. Eventually it may include the following items.

Estimating Concrete Wall Checklist

- ❑ Formwork (sfca)
- ❑ Concrete (cy)
- ❑ Rebar (lb ro ton)
- ❑ Rigid insulation (sf) and damp proofing (sf)
- ❑ Waterstop and keyway (lf)
- ❑ Control joint and chamfer strip (lf)
- ❑ Patching, rubbing, and finishing (sf), if exposed
- ❑ Excavation, backfill, and compact (cy)

Estimating Math for Formwork and Concrete

Wall Area = Wall Length × Wall Height
Formwork = (Wall Length + Wall Thickness) × 2 × Wall Height
Concrete = Wall Length × Wall Thickness × Wall Height

Estimating Example 1

Wall is 40 ft long, 10 inches thick, and 8 ft high from top of footing

Wall area = 40 × 8 = 320 sf
Formwork = (40 + 0.83) × 2 × 8 = 654 sfca
Concrete = 40 × 0.83 × 8 = 266 cf (that is, 266/27 = 9.9 cy)

Allow 10% waste, concrete volume is 9.9 × (1 + 10%) = 10.9 cy.

For some retaining walls, heights change due to the varying ground elevations (and with stepped footings). To take it off, you can divide wall lengths into several segments at each elevation change, and take an average height.

Estimating Example 2

A 20 ft long, 8 inch thick concrete wall is shown with two contour lines. Grade elevation is 100 ft. Top elevation of wall footing is 98 ft. The contour lines divide the wall into three segments. Segments #1 and #3 are 5 ft long and segment #2 is 10 ft long. Assume the wall elevations change gradually in a linear pattern.

Between two contour lines (segment #2), the wall top elevation increases by 103 ft − 101 ft = 2 ft.

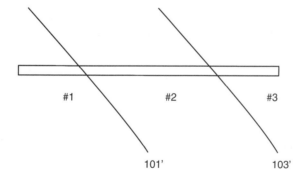

This segment is 10 ft, so wall elevation change per length = 2/10 = 0.2 ft.

This means for every running foot of wall, elevation changes by 0.2 ft.

So the beginning elevation for the whole wall (the left end of segment #1) is 101 ft − 5 ft × 0.2 = 100 ft.

The ending elevation for the whole wall (the right end of segment #3) is 103 ft + 5 ft × 0.2 = 104 ft.

The average elevation for entire wall length is (100 ft + 104 ft)/2 = 102 ft.

Therefore, the average wall height is 102 ft − 98 ft = 4 ft (including below grade).

$$\text{Total wall area} = 20 \times 4 = 80 \text{ sf (aboveground area is } 20 \times 2 = 40 \text{ sf)}$$
$$\text{Formwork} = (20 + 0.67) \times 2 \times 4 = 166 \text{ sfca}$$
$$\text{Concrete} = 20 \times 0.67 \times 4 = 53.6 \text{ CF, i.e. } 53.6/27 = 2 \text{ CY}$$

Allow 10% waste, concrete volume is $2 \times (1 + 10\%) = 2.2$ cy.

Alternatively, you can calculate by each segment (with the same results):

Segment	Start Elev.	End Elev.	Average Elev.	Wall Height
#1	100'	101'	100.5'	2.5'
#2	101'	103'	102'	4'
#3	103'	104'	103.5'	5.5'

Formwork for segment #1 = $5 \times 2 \times 2.5 = 25$ sfca

Formwork for segment #2 = $10 \times 2 \times 4 = 80$ sfca

Formwork for segment #3 = $5 \times 2 \times 5.5 = 55$ sfca

Formwork for wall width at ends = $0.67 \times 2.5 + 0.67 \times 5.5 = 6$ sfca

Total formwork = $25 + 80 + 55 + 6 = 166$ sfca

Concrete for segment 1 = $5 \times 0.67 \times 2.5 = 8.4$ cf

Concrete for segment 2 = $10 \times 0.67 \times 4 = 26.8$ cf

Concrete for segment 3 = $5 \times 0.67 \times 5.5 = 18.4$ cf

Total concrete = $8.4 + 26.8 + 18.4 = 53.6$ cf (that is, $53.6/27 = 2$ cy)

Allow 10% waste, concrete volume is $2 \times (1 + 10\%) = 2.2$ cy.

ESTIMATING CONCRETE PAD FOOTINGS

Concrete pad footings are isolated foundations under columns. Normally, the takeoff starts with counting the number of pads and includes:

Estimating Concrete Pad Footings Checklist

- Pad layout (l/s)
- Excavation (cy)
- Fine grading (sf)
- Formwork (sfca)
- Concrete (cy)
- Rebar (lb or ton)
- Embeds (ea) (e.g., anchor bolts and bearing plates)
- Backfill (cy)

Download this form at www.DEWALT.com/guides

Estimating Math for Formwork and Concrete

$$\text{Formwork} = \text{Number of Pads} \times (\text{Pad Length} + \text{Pad Width}) \times 2 \times \text{Pad Depth}$$
$$\text{Concrete} = \text{Number of Pads} \times \text{Pad Length} \times \text{Pad Width} \times \text{Pad Depth}$$

Estimating Example

Pad is 1.5 ft long, 1.5 ft wide, and 12 inches deep; quantity: four.

$$\text{Formwork} = 4 \times (1.5 + 1.5) \times 2 \times 1 = 24 \text{ sfca}$$
$$\text{Concrete} = 4 \times 1.5 \times 1.5 \times 1 = 9 \text{ cf (that is, } 9/27 = 0.34 \text{ cy)}$$

Allow 10% waste, concrete volume is $0.34 \times (1 + 10\%) = 0.38$ cy.

ESTIMATING CONCRETE SLABS

Concrete slabs can be as simple as a driveway for a residential home or as complicated as a suspended slab and band in a high-rise office tower. Regardless, the takeoff always begins with measuring slab areas and may include:

Estimating Concrete Slabs Checklist

- ❏ Granular base (sf or cy)
- ❏ Fine grading (sf)
- ❏ Formwork (sfca)
- ❏ Concrete (cy)
- ❏ Rebar (lb) or wire mesh (sf or rolls)
- ❏ Poly vapor barrier (sf)
- ❏ Saw-cut control joint (lf)
- ❏ Curing and sealing (sf, gallon or pail)
- ❏ Hardener (sf, gallon or pail)
- ❏ Rigid insulation (sf or rolls)
- ❏ Finishing (sf) (e.g., trowel, broom)
- ❏ Scaffolding and shoring for suspended slab formwork (l/s)

Download this form at **www.DeWALT.com/guides**

Estimating Math for Rectangle Slab on Grade

$$\text{Slab Area} = \text{Slab Length} \times \text{Slab Width}$$
$$\text{Edge Formwork} = (\text{Slab Length} + \text{Slab Width}) \times 2 \times \text{Slab Thickness}$$
$$\text{Concrete} = \text{Slab Length} \times \text{Slab Width} \times \text{Slab Thickness}$$

Estimating Example 1

A driveway slab is 20 ft wide, 24 ft long, and 4 inches thick.

$$\text{Formwork} = (20 + 24) \times 2 \times 0.33 = 29 \text{ sfca}$$
$$\text{Concrete} = 20 \times 24 \times 0.33 = 160 \text{ cf (that is, } 160/27 = 5.9 \text{ cy)}$$

Allow 10% waste, concrete volume is $5.9 \times (1 + 10\%) = 6.5$ cy.

Note that edge formwork is not always required for slabs, especially when the slab goes against a wall. For irregular shapes of slabs, divide them into smaller sections for easy calculations.

Slabs frequently are thickened under a load-bearing wall or at the edge. These thickenings (or slab turn-downs) should be adequately accounted for.

Estimating Example 2

An exterior slab edge of 100 ft long is shown here:

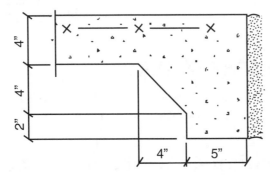

$$\text{Edge Formwork} = 100 \text{ ft} \times (4 \text{ inches} + 4 \text{ inches} + 2 \text{ inches})/12 = 84 \text{ sfca}$$

The 4 inch concrete above the edge should already been included in the main slab.

$$\text{Edge Concrete} = 100 \text{ ft} \times (4 \text{ inches} + 5 \text{ inches})/12 \times (2 \text{ inches} + 4 \text{ inches})/12$$
$$= 37.5 \text{ cf (that is, } 37.5/27 = 1.4 \text{ cy)}$$

Allow 10% waste, concrete volume is $1.4 \times (1 + 10\%) = 1.54$ cy.
Please allow enough concrete materials, keeping in mind your total is an estimate only.

ESTIMATING CONCRETE COLUMNS

Concrete columns are vertical supporting structural members. They can be found in both residential and commercial/institutional projects to support parking facilities, exterior canopies and main roof decks. The takeoff starts with counting the number of columns according to their dimensions.

Estimating Math for Rectangular Columns

$$\text{Formwork} = \text{Count of Columns} \times (\text{Section Length} + \text{Section Width}) \times 2 \times \text{Column Height}$$
$$\text{Concrete} = \text{Count of Columns} \times \text{Section Length} \times \text{Section Width} \times \text{Column Height}$$

Estimating Example

For 12 ea 24 inch × 24 inch square column, 15 ft high

$$\text{Formwork} = 12 \times (2 + 2) \times 2 \times 15 = 1{,}440 \text{ sfca}$$
$$\text{Concrete} = 12 \times 2 \times 2 \times 15 = 720 \text{ cf (that is, } 720/27 = 26.7 \text{ cy)}$$

Allow 10% waste, concrete volume is $26.7 \times (1 + 10\%) = 29.4$ cy.
If columns are totally exposed, Finishing Area = Formwork = 1,440 sf

$$\text{Chamfer Strip} = \text{Count of Columns} \times \text{Number of Corners} \times \text{Column Height}$$
$$= 12 \times 4 \times 15 = 720 \text{ lf}$$

ESTIMATING CONCRETE TIE BEAMS

Concrete tie beams are more common in commercial and institutional projects. Suspended slab bands are also sometimes called "beams" and they are estimated in a similar way.

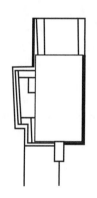
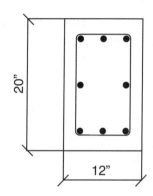

20"

12"

CB1
(6) #8 [3 TOP, 3 BTM.], (2)#7
MID, 2" MIN CLR., #4 STIRRUPS @
8" O.C., STD. HOOK ON T&B BARS
INTO COL. OUTSIDE VERT. REINF.

Estimating Math

$$\text{Side Formwork} = \text{Beam Height} \times \text{Beam Length} \times 2$$
$$\text{Bottom Formwork} = \text{Beam Width} \times \text{Beam Length}$$
$$\text{Concrete} = \text{Beam Height} \times \text{Beam Width} \times \text{Beam Length}$$

Estimating Example

For a 20 inch $\times$ 12 inch beam, 30 ft long:

$$\text{Side Formwork} = 1.67 \text{ ft} \times 30 \text{ ft} \times 2 = 100 \text{ sfca}$$
$$\text{Bottom Formwork} = 1 \text{ ft} \times 30 \text{ ft} = 30 \text{ sfca (not required if it sits on top of a wall)}$$
$$\text{Total Formwork} = 100 + 30 = 130 \text{ sfca}$$
$$\text{Concrete} = 1.67 \times 1 \times 30 = 50.1 \text{ cf (that is, } 50.1/27 = 1.9 \text{ cy)}$$

Allow 15% waste, concrete volume is $1.9 \times (1 + 15\%) = 2.2$ cy.
Check to see if finishing is required for both sides and bottom (130 sf).

ESTIMATING EXTERIOR CONCRETE WORK

Exterior concrete work shown on site or landscaping drawings is often overlooked, but it can be quite costly. The following list includes some common items for takeoff.

Estimating Exterior Concrete Work Checklist

- ❏ Concrete sidewalks adjacent to the building
- ❏ Concrete sidewalks far away from the building (or even on the road)
- ❏ Driveway aprons
- ❏ Dumpster pad and apron
- ❏ Transformer or equipment pads
- ❏ Exit stairways and landing
- ❏ Concrete retaining wall
- ❏ Foundation for masonry screen wall
- ❏ Foundation for bollards, signage, and light pole bases
- ❏ Foundation for site features such as gazebos, fountains, arbors, and fences
- ❏ Concrete encasement for utility pipes and electrical conduits

Download this form at **www.DeWALT.com/guides**

It is important to be meticulous during takeoff of these items, rather than plugging in a casual cash allowance. Devote serious estimating time to this portion of the project. Exterior work is normally done by the end of the project, but you should try to cover these costs from the beginning.

ESTIMATING REINFORCING

Reinforcing is required for concrete and masonry work. Rebar is designated by bar size, that is, the nominal diameter of the bar in eighths of an inch. For example, a No. 6 bar has a nominal diameter of 6/8 inch or 3/4 inch. Rebar is taken off by the linear foot and the number of pieces. On large commercial jobs, the total lengths are then converted to weights.

Estimating formula for rebar weight conversion is:

$$\text{Bar Weight (lb)} = \text{Number of Bars (ea)} \times \text{Bar Length (ft)} \times \text{Unit Weight (lb/ft)}$$

The following table gives the reinforcing steel weight information, according to bar sizes.

Bar Size	Weight (lb/ft)
#2	0.167
#3	0.376
#4	0.668
#5	1.043
#6	1.502
#7	2.044
#8	2.670
#9	3.400
#10	4.303

Wire mesh is also furnished by rebar suppliers to use in slabs. It is estimated in square feet, then converted in rolls.

Estimating Example 1

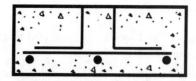

A 200 ft long × 2.5 ft wide × 12 inch high strip footing is reinforced as:
    4 ea # 5 bars continuous, 1 ea # 5 transverse bars @ 48 inches on center
    2 ea # 7 dowels to masonry wall above @ 48 inches on center (assume 3.5 ft long including the hook length)

Continuous Bar = Footing Length × Count of Bars = 200 × 4 = 800 ft of # 5 bars
Transverse Bar = (Footing Length/Bar Spacing + 1) × Footing Width × Count of Bars
= (200/4 + 1) × 2.5 × 1 = 128 ft of # 5 bars
Dowels = (Footing Length/Dowel Spacing + 1) × Dowel Length × Count of Dowels
= (200/4 + 1) × 3.5 × 2 = 357 ft of # 7 bars

So there is net length of 800 ft + 128 ft = 928 ft of # 5 bars and 357 ft of # 7 bars.
Bars normally are bought in 20 ft pieces, so use lap of 2.5 ft for # 5 bars and 3.5 ft for # 7 bars (i.e., 48 inch bar diameter).

Total 20 ft pieces of # 5 bars are 928/(20 − 2.5) = 54 ea.
Total 20 ft pieces of # 7 bars are 357/(20 − 3.5) = 22 ea.
The total weight of bars is 54 ea × 20 ft × 1.043 lb/ft + 22 ea × 20 ft × 2.044 lb/ft = 2,026 lb.

Allow 20% for waste, 2,026 × (1 + 20%) = 2,432 lb, or 2,432/2,000 = 1.22 tons.

Estimating Example 2

10 ea 3 ft × 3 ft × 12 inch pad footing reinforced with two # 5 bars at bottom each way.

Rebar = Count of Pads × (Pad Length + Pad Width) × Count of Bars
= 10 × (3 ft + 3 ft) × 2 = 120 ft of # 5 bars

Bars normally are bought in 20 ft piece, so use lap of 2.5 ft for # 5 bars (i.e., 48 inch bar diameter).

Total 20 ft pieces of # 5 bars are 120/(20 − 2.5) = 7 ea.
Weight is 7 ea × 20 ft × 1.043 lb/ft = 147 lb.

Allow 20% for waste, 147 × (1 + 20%) = 177 lb, or 177/2,000 = 0.09 ton.

Estimating Example 3

A 200 ft × 80 ft slab on grade reinforced with wire mesh lapped 6 inches:
    Wire mesh is bought in 150 ft × 5 ft rolls.

Roll Effective Area = (150 − 0.5) × (5 − 0.5) = 672 sf
Rolls = Slab Width × Slab Length/Roll Effective Area = 200 × 80/672 = 24 ea

Allow 10% for waste, 24 × (1 + 10%) = 27 rolls.

ESTIMATING MASONRY

Following is a systematic approach to takeoff masonry work.

- Identify material requirements, such as block/brick type and manufacturer; rebar lap; cell-fill concrete strength; block fire resistance requirements.

- From foundation plan, takeoff stem wall along the building perimeter.

- Compare roof framing plan with exterior elevations for actual wall height; takeoff main exterior wall above grade.

- Find masonry bond beams on roofs and parapets and over openings.

- Add half blocks around openings, at tie columns, and at wall control joints.

- Figure corner blocks as necessary.

- Include building interior elevator/mechanical/electrical shaft walls.

- Include building exterior veneer and columns (e.g., split face blocks, face bricks, cultured or cast stone).

- From site plan, find dumpsters and screen walls.

- Takeoff structural precast lintels per their dimensions.

- Takeoff architectural precast items such as column caps, planter wall caps, copings, water tables, window sills.

- Count accessories such as hollow metal door frames in masonry wall.

- Get quote on special material such as custom bricks and stone veneer.

- Get quote on reinforcing steel for masonry dowels.

- Get quote for equipment rental of scaffolding, wall bracing.

Estimating Masonry Checklist

- ❏ Blocks (ea), by type and grade (e.g., regular gray, split face, 8 inch, 12 inch)
- ❏ Bricks (ea)
- ❏ Stone veneer (sf)
- ❏ Rebar (lb or ton)
- ❏ Cell fill concrete (cy)
- ❏ Mortar (bags)
- ❏ Sand (cy or ton)
- ❏ Scaffolding (sf)
- ❏ Wall bracing (sf of l/s)
- ❏ Control joint filler (lf)
- ❏ Joint reinforcement (lf)
- ❏ Flashing (lf)
- ❏ Wall tie (ea)
- ❏ Anchors (ea)
- ❏ Weep hole (ea)
- ❏ Bearing angles, channels, or plates (ea or lf)
- ❏ Foam or rigid insulation (sf or bags)
- ❏ Firestopping (sf)
- ❏ Installing door frames (ea)
- ❏ Precast lintels and sills (lf or ea)
- ❏ Wall cleaning (sf)

Download this form at **www.DeWALT.com/guides**

Estimating Math

For the sake of simplicity, use the following rules of thumb.

Item	Quantities (Net)
Blocks & Bricks	
Standard Block	1.125 blocks per sf of wall area
Half Block	2.25 blocks per sf of wall area
Face Brick Modular	7.0 bricks per sf of wall area
Oversize Brick	6.0 bricks per sf of wall area
Utility Brick	3.0 bricks per sf of wall area

Item	Quantities (Net)
Mortar	
Block	3 bags per 100 block
Face Brick Modular	7 bags per 1,000 brick
Oversized Brick	8 bags per 1,000 brick
Utility Brick	10 bags per 1,000 brick
Sand	1 cy per 7 bags mortar
Cell-fill Concrete (Grout)	
6×8×16	0.17 cf/block
8×8×16	0.25 cf/block
10×8×16	0.33 cf/block
12×8×16	0.39 cf/block
6×8×16 Bond Beam	0.173 cf per lf
8×8×16 Bond Beam	0.22 cf per lf
8×8×16 Deep Bond Beam	0.46 cf per lf
12×8×16 Bond Beam	0.37 cf per lf
12×8×16 Deep Bond Beam	0.74 cf per lf

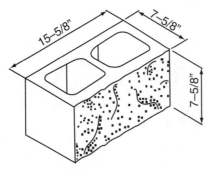

In estimating masonry, the essential job is to determine the quantity of blocks or bricks needed. Everything else will be based on that figure.

Estimating Example

An 8 inch block wall is 100 ft long and 3 ft long.

$$\text{Wall area} = 100 \times 3 = 300 \text{ sf}$$

From the previous table:

1.125 blocks per sf of wall area: Number of Blocks = 300 × 1.125 = 338 ea

3 bags of mortar per 100 blocks: Mortar = 3 × 338/100 = 11 bags

1 cy sand per 7 bags of mortar: Sand = 1 × 11/7 = 1.6 cy

0.25 cf grout per 8 inch block (not including bond beams): Cell Fill Grout = 338 × 0.25 = 85 cf
Note all quantities above are net. Please add waste.

ESTIMATING STRUCTURAL STEEL

Estimating Structural Steel Checklist

❏ Open web steel joists (lf converted to tons)

❏ Metal roof and floor decking (sf converted to tons, note details such as type, gage, thickness, finish, method of attachment)

❏ Structural steel shapes (lf or ea converted to lb, then tons)

    ❏ Beams and girders

    ❏ Columns, tubing, and pipes

    ❏ Angles and channels

    ❏ Plates and connectors

    ❏ Joist bridging and girts

    ❏ Hanger rods

    ❏ Wall bracing and end anchorage

❏ Anchor bolts and base plates (ea)

❏ Waste and connection (l/s)

❏ Metal canopy (ea or sf)

❏ Metal stairs (ea)

❏ Roof or elevator pit access ladder (ea)

❏ Elevator hoist or divider beams (ea)

❏ Metal gates and frames (ea)

❏ Embedded framing angle or plates (lf)

    ❏ Foundation support angle

    ❏ Masonry veneer support angle and wall top angle

    ❏ Roof parapet angle

    ❏ RTU frames

    ❏ Overhead door and sliding door framing

❏ Protective covers (ea)

    ❏ Column corners guards

    ❏ Trench drain and catch basin covers

    ❏ Expansion joint covers

    ❏ Thermostat covers

    ❏ Galvanized bollards

❏ Support framing for hot water tanks, heaters, etc. (ea or lf)

❏ Handrail and guardrail at stairs, walls, and balconies (lf)

❏ Paint or prime (l/s, sf, or lf)

Download this form at **www.DeWALT.com/guides**

Estimating Structural Shapes and Joists

Take the following steps

1. Divide the building into smaller portions of floor areas or bays.
2. Identify different shapes and section sizes.
3. Measure and sum up lengths for each shape separately.
4. Multiply lengths by the unit weight per linear foot for each shape.
5. Sum up the total weight, figure connection and waste (10% to 20%).
6. Covert the result to tons.

Estimating Example

Calculate the total net weight of the following shapes.

Designation	Length	Count
W 12 × 14	15' 6-5/8"	4
S 5 × 10	20' 7-3/4"	3
HSS 20 × 20 × 3/8	25' 2-3/8"	2

Calculations:

Linear feet of W12 × 14 = 4 × 15' 6-5/8" = 4 × 15.55' = 62.20'

Linear feet of S5 × 10 = 3 × 20' 7-3/4" = 3 × 20.64' = 61.92'

Linear feet of HSS 20 × 20 × 3/8 = 2 × 25' 2-3/8" = 2 × 25.20' = 50.40'

Look up their unit weights (pounds per linear foot) from the weight tables:

Pounds of W12 × 14 = 62.20' × 14 lb/ft = 871 lb

Pounds of S5 × 10 = 61.92' × 10 lb/ft = 620 lb

Pounds of HSS 20 × 20 × 3/8 = 50.40' × 103.22 lb/ft = 5,203 lb

Total Weight: 871 + 620 + 5,203 = 6,693 lb (net quantity)

Convert to tons (short ton): 6,693/2,000 = 3.35 tons

ESTIMATING ROUGH CARPENTRY AND WOOD FRAMING

Residential framing work normally consists of the following:

- Floor Framing
  - Sill plates and nailers
  - Posts, girders, and column base connectors
  - Joists and beams including rim joists, joist cross bracing, hangers
  - Trimmers and headers for floor opening
  - Plywood floor sheathing
  - Wood decks

- Wall Framing
  - Wall studs, plates, and shear wall anchors
  - Trimmers and headers for wall opening
  - Plywood wall sheathing
  - Blocking for floor connections and wall bracings
  - Wood stairs (prefabricated or framed)
- Roof/Ceiling Framing
  - Truss and joists with backing
  - Rafters, ridges, beams, collar ties, rat runs
  - Plywood roof sheathing
  - Plank decking and strapping
  - Fascia, soffit, overhang, balcony, and terrace

Back framing work consists of the following (framers usually exclude these):

- Roof blocking, including locations at canopy, parapet, sidewall, cant strip
- Roof opening blocking, such as for mechanical curb, roof hatch, and skylights
- Wall-hung equipment backing such as ladders and electrical panels
- Door bucks and window blocking
- Millwork and toilet accessories backing
- Wood wainscot or sanitary base
- Exterior and interior trim backing
- Exterior signage support

The phrase "framing material" could mean any of the following:

- TJI wood joists and beams
- Preengineered wood trusses
- Laminated beams, columns, and arches
- Dimensional lumber and plywood

For the first three items, you will get quotes from manufacturers who specialize on these prefabricated systems, and then add your own labor and equipment. For rough lumber and plywood, however, you will take off quantities on your own and then call the lumberyards. Suppliers can figure out quantities for you, but they may be not accurate enough.

ESTIMATING DIMENSIONAL LUMBER

Lumbers are measured by their length (lf) and converted to board feet (bf). It is important to separate lumbers by different grades (e.g. kiln dry, pressured treated, fire rated). For takeoff, you need to make detailed notes for each item.

Estimating Math

$$\text{Board Foot} = \text{Lumber Pieces} \times \text{Section Thickness} \times \text{Section Width}/12 \times \text{Length}$$

Estimating Example

For 10 pieces of 14 ft long 2 × 4 lumbers:

$$\text{Total Board Feet} = 10 \times (2 \times 4)/12 \times 14 = 94 \text{ bf}$$

Some estimators memorize conversion factors such as 0.67 bf/lf for 2 × 4 lumbers and 1.00 bf/lf for 2 × 6 lumbers, and so forth. The following table gives a list of conversion factors. For the previous example, total board feet = 10 × 14 lf × 0.67 bf/lf = 94 bf. The result is same.

When estimating residential framing lumber, it is common to allow 25% or more for the waste, especially when drawings are not clear. For the previous example, use 94 × (1 + 25%) = 118 bf of lumber.

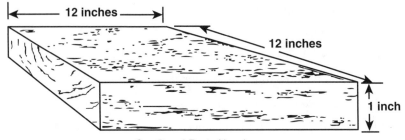

One Board-Foot of Lumber

Nominal Size (in × in)	Actual Size (in × in)	Board Feet (bf) per Linear Foot (lf) of lumber
1 × 2	3/4 × 1-1/2	0.17
1 × 3	3/4 × 2-1/2	0.25
1 × 4	3/4 × 3-1/2	0.33
1 × 6	3/4 × 5-1/2	0.50
1 × 8	3/4 × 7-1/4	0.67
1 × 10	3/4 × 9-1/4	0.83
1 × 12	3/4 × 11-1/4	1.00
2 × 2	1-1/2 × 1-1/2	0.33
2 × 3	1-1/2 × 2-1/2	0.50
2 × 4	1-1/2 × 3-1/2	0.67
2 × 6	1-1/2 × 5-1/2	1.00
2 × 8	1-1/2 × 7-1/4	1.33
2 × 10	1-1/2 × 9-1/4	1.67
2 × 12	1-1/2 × 11-1/4	2.00
2 × 14	1-1/2 × 13-1/4	2.33
4 × 4	3-1/2 × 3-1/2	1.33
6 × 6	5-1/2 × 5-1/2	3.00
8 × 8	7-1/2 × 7-1/2	5.33

ESTIMATING PLYWOOD SHEETS

Plywood is used as sheathings on floor, wall, and roof. They are measured by the area to be covered (sf) and converted to sheets (ea). The most common sheet size is 4 ft × 8 ft, although other dimensions are also available. It is also important to note the plywood grades (e.g., CDX, fire rated, OSB) and thickness (e.g., ½ inch, 5/8 inch, ¾ inch) for pricing purposes.

Estimating Math

$$\text{Plywood Sheets} = \text{Coverage Area}/32$$

It is common to allow 15% or more for the waste. Remember to round up the calculation results to whole numbers of sheets.

Estimating Example

If the floor area is 35 ft × 24 ft, then the number of 4 × 8 plywood sheets is:

$$35 \times 24/32 = 26.25 \text{ sheets}$$

Allow 15% waste, 26.25 × (1 + 15%) = 30.2 sheets, so round up to 31 sheets.

ESTIMATING FLOOR JOISTS AND ROOF TRUSSES

Framers need to figure out the labor and equipment to install roof trusses and floor joists. They are counted by each, based on their spacing. If they are preengineered, the actual quantities can be verified with the truss/joist suppliers. If they are framed (using 2 × 8 lumber, etc), then you must figure out the actual board feet of lumber.

Estimating Math

$$\text{Number of Trusses/Joists} = \text{Running Length}/\text{Spacing} + 1$$

Base the calculation on the same unit (foot or inch), and round up the result to the next whole number. In some situations (e.g., where partitions run parallel to joist), the calculated result may not be enough. Additional quantities should be allowed.

Estimating Example 1

The roof is 150 ft long and trusses are placed 24 inches (i.e., 2 ft) on center, then the number of trusses is: 150/2 + 1 = 75 + 1 = 76 ea.

Estimating Example 2

The floor is 135 ft long and joists are placed 16 inches (i.e., 1.33 ft) on center, then the number of joists is: 135/1.33 + 1 = 102.5 or 103 ea.

ESTIMATING WALL STUDS

In theory, the number of wall studs can be figured in the same way as trusses and joists, with a minor difference. Today, more floor and roof systems are preengineered, but walls are still being framed in the field with huge waste. (Prefabricated wall panel systems have yet to become an industry norm.) Thus, if you are framing the wall, allow for enough lumber to complete the project.

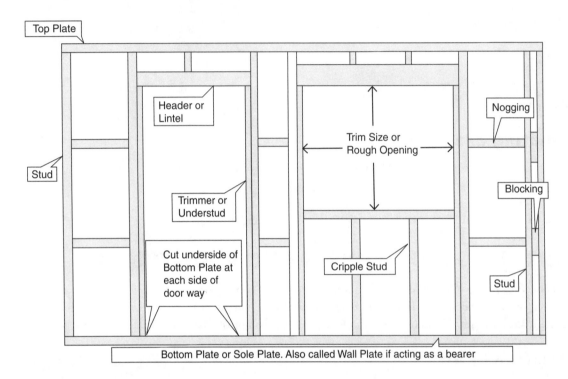

Bottom Plate or Sole Plate. Also called Wall Plate if acting as a bearer

Estimating Math

For wall stud spacing 12 inches or less:

$$\text{Number of Wall Studs} = \text{Wall Length/Stud Spacing} + 1$$

For wall stud spacing greater than 12 inches:

$$\text{Number of Wall Studs} = \text{Wall Length (ft) (one stud per one ft of wall)}$$
$$\text{Total Board Feet} = \text{Number of Wall Studs} \times \text{Wall Height} \times \text{Board Foot Conversion Factor}$$

Estimating Example 1

A 25 ft long × 10 ft high wall, where 2 × 4 studs are spaced 12 inches on center; normally, the math is:

Stud Spacing	Multiply Partition Length (ft) by	Add
12″	1.00	1
16″	0.75	1
20″	0.60	1
24″	0.50	1

There are $25 \times 1 + 1 = 26$ ea wood studs.

$$\text{Total Board Feet} = 26 \text{ ea} \times 10 \text{ lf} \times 0.67 \text{ bf/lf} = 175 \text{ bf}$$

Allow 25% waste, you need at least $175 \times (1 + 25\%) = 219$ bf for studs

Estimating Example 2

A 150 ft long $\times$ 12 ft high wall, where 2×6 studs are spaced 16 inches on center; by math, there are $150 \times 0.75 + 1 = 113.5$ or 114 ea wall studs. However, simply allow one stud per foot of wall, or 150 ea wall studs, to ensure adequate materials to frame the wall.

$$\text{Total Board Feet} = 150 \text{ ea} \times 12 \text{ lf} \times 1.00 \text{ bf/lf} = 1{,}800 \text{ bf}$$

Allow 25% waste, you need at least $1{,}800 \times (1 + 25\%) = 2{,}250$ bf for studs

ESTIMATING ROOF RAFTERS

The calculation of roof rafter requires understanding the concept of roof slope, which is the ratio of rise over run (e.g., 4:12, 5:12). Roof pitch is half of roof slope. The term *roof slope* is a more common term than *roof pitch*. Please include the eaves and overhang distance when measuring run or rise.

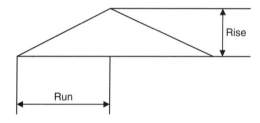

Estimating Math

$$\text{Span} = 2 \times \text{Run}$$
$$\text{Pitch} = \text{Rise/Span}$$
$$\text{Slope} = \text{Rise/Run}$$

$$\text{Total Run} = \text{Run Distance} + \text{Eavehang Distance}$$
$$\text{Common Rafter Length} = \text{Total Run} \times \text{Multiplication Factor 1}$$
$$\text{Hip/Valley Rafter Length} = \text{Total Run} \times \text{Multiplication Factor 2}$$

(See the multiplication factor tables as follows.)

Multiplication Factors for Sloped Roof Rafter

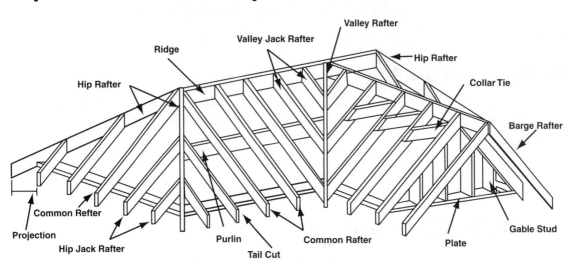

Roof Pitch	Roof Slope	Multiply Length of Run By	
		For Common Rafter	For Hip/Valley Rafter
1/12	2 in 12	1.014	1.424
1/8	3 in 12	1.031	1.436
1/6	4 in 12	1.054	1.453
5/24	5 in 12	1.083	1.474
1/4	6 in 12	1.118	1.500
7/24	7 in 12	1.158	1.530
1/3	8 in 12	1.202	1.564
3/8	9 in 12	1.250	1.601
5/12	10 in 12	1.302	1.642
11/24	11 in 12	1.357	1.685
1/2	12 in 12	1.413	1.732
13/24	13 in 12	1.474	1.782
7/12	14 in 12	1.537	1.833
5/8	15 in 12	1.601	1.887
2/3	16 in 12	1.667	1.944
17/24	17 in 12	1.734	2.002
3/4	18 in 12	1.803	2.032
19/24	19 in 12	1.875	2.123
5/6	20 in 12	1.948	2.186
7/8	21 in 12	2.010	2.250
11/12	22 in 12	2.083	2.315
23/24	23 in 12	2.167	2.382
Full	24 in 12	2.240	2.450

Estimating Example

Roof slope is 6/12, run is 14.5 ft, overhang is 1.5 ft:
Effective Run Distance = Run + Overhang = 14.5 + 1.5 = 16 ft
From previous table, Common Rafter Length = 16 ft × 1.118 = 17.9 ft

NOTE

If you want to be more precise, deduct half of the ridge board thickness from the run distance. For example, if ridge board is 2 × 6, run + overhang is 16 ft or 192 inches, then deduct half of 1.5 inches: 192 inches − 1/2 × 1.5 inches = 191.25 inches. Then common rafter length will be 191.25 inches × 1.118 = 213′ − 13/16″.

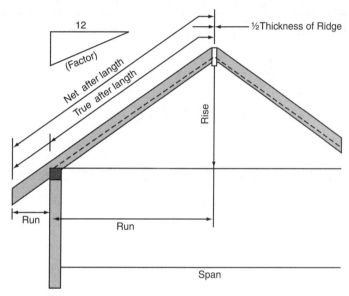

FRAMING MATERIAL RULES OF THUMB

Wall Framing

Stud Size	Spacing	Board Foot per Square Foot of Wall
2 × 4	12″	1.27
	16″	1.17
	20″	1.10
	24″	1.07
	Staggered	1.69
2 × 6	16″	1.51
	20″	1.44
	24″	1.38

Estimating Example

A 120 ft long, 8 ft high 2 × 6 wood framed wall stud spacing 16 inches o.c., then the wall area is 120 × 8 = 960 sf.

Using the previous table, the framing lumber needed is approximately 960 × 1.51 = 1,450 bf, including studs, plates, blocking for corners and openings, etc.

Floor/Ceiling Framing

Joist Size	Spacing	Board Foot per Square Foot of Floor/Ceiling
2 × 6	12″	1.28
	16″	1.02
	24″	0.78
2 × 8	12″	1.71
	16″	1.36
	24″	1.03
2 × 10	12″	2.14
	16″	1.71
	24″	1.30
2 × 12	12″	2.56
	16″	2.05
	24″	1.56

ESTIMATING FINISH CARPENTRY

The phrase "finish carpentry" can have various meanings. The following takeoff checklist serves as a quick reference.

Estimating Finish Carpentry Checklist

- ❏ Residential casework (lf or ea) such as kitchen and bathroom cabinets
- ❏ Solid surface countertops (sf) such as stone or granite
- ❏ Custom millwork (lf or ea) including cabinets, shelving, stairs, railing
- ❏ Wood trims (lf)
    - ❏ Exterior wood trims (lf)
    - ❏ Interior wood baseboards (lf)
    - ❏ Cornice trims (lf)
    - ❏ Crown moldings (lf)
    - ❏ Window sills (lf)
    - ❏ Door and wall opening casings (lf)
    - ❏ Corridor wood handrail (lf)
- ❏ Exterior vinyl soffit (sf)
- ❏ FRP panels (sf)
- ❏ Wood wall paneling or wainscot (sf)
- ❏ Labor-only package
    - ❏ Installation of doors, frames, and hardware (ea)
    - ❏ Installation of vinyl windows (ea)
    - ❏ Installation of washroom accessories (ea)
    - ❏ Installation of millwork and FFE (ea)

ESTIMATING ROOFING AND SIDING

Use the following checklist for roofing/siding takeoff (with measurement units).

Estimating Roofing and Siding Checklist

- ❏ Flat roof (sf converted to sq) such as SBS, single-ply, EPDM
- ❏ Sloped roof (sf converted to sq) such as shingles, tile, or standing seam
- ❏ Felts, sheathing paper, and underlayment (sf)
- ❏ Roof insulation (sf)
- ❏ Vapor barrier (sf)
- ❏ Down sprouts and gutters (lf)
- ❏ Scuppers and roof drains (lf)
- ❏ Flashing, expansion joints, and gravel stops (lf)
- ❏ Louvers and vents (lf or ea)
- ❏ Roof pavers (sf)
- ❏ Coping (lf)
- ❏ Ridge strips (lf)
- ❏ Fascias (lf)
- ❏ Reglets (lf)
- ❏ Sidings (sf), including vinyl, wood, and metal sidings
- ❏ Metal or vinyl soffits (sf)
- ❏ Parapet rain-lock panels (sf)
- ❏ Skylights (ea)
- ❏ Roof hatch with posts (ea)

Download this form at **www.DEWALT.com/guides**

ESTIMATING SLOPED ROOF AREA

To estimate the actual sloped roof area, simply measure the flat projection area (as shown on drawings), and then multiply the factor below accordingly.

Roof Pitch	Roof Slope	Multiply Flat Roof Area By
1/12	2 in 12	1.014
1/8	3 in 12	1.031
1/6	4 in 12	1.054
5/24	5 in 12	1.083
¼	6 in 12	1.118
7/24	7 in 12	1.158
1/3	8 in 12	1.202

Roof Pitch	Roof Slope	Multiply Flat Roof Area By
$\frac{3}{8}$	9 in 12	1.250
$\frac{5}{12}$	10 in 12	1.302
$\frac{11}{24}$	11 in 12	1.357
$\frac{1}{2}$	12 in 12	1.413
$\frac{13}{24}$	13 in 12	1.474
$\frac{7}{12}$	14 in 12	1.537
$\frac{5}{8}$	15 in 12	1.601
$\frac{2}{3}$	16 in 12	1.667
$\frac{17}{24}$	17 in 12	1.734
$\frac{3}{4}$	18 in 12	1.803
$\frac{19}{24}$	19 in 12	1.875
$\frac{5}{6}$	20 in 12	1.948
$\frac{7}{8}$	21 in 12	2.010
$\frac{11}{22}$	22 in 12	2.083
$\frac{23}{24}$	23 in 12	2.167
Full	24 in 12	2.240

Estimating Example

A roof is sloped as 5:12. It is measured as 100 ft × 70 ft on drawings.

$$\text{Flat area} = 100 \text{ ft} \times 70 \text{ ft} = 7,000 \text{ sf}$$
$$\text{Actual area} = 7,000 \times 1.083 = 7,581 \text{ sf (net quantity)}$$

ESTIMATING MISCELLANEOUS DIVISION 7 ITEMS

Read specifications for sections of Division 7, "Thermal and Moisture Protection," carefully examine drawings to define the project's scope. The following list includes common Division 7 items.

- Batt insulation (sf)
- Spray foam insulation (sf)
- Rigid insulation (sf) for walls and slab edges
- Protection board (sf)
- Elevator pit waterproofing (sf)
- Patio waterproof membrane (sf)
- Foundation wall damp proofing (sf)
- Balcony deck coating (sf)
- Vapor/air barrier (sf)
- Poly underslab on grade (sf)
- Blueskins to windows (lf, sf, or ea)

- Firestopping (sf)
- Spray fireproofing (sf)
- Caulking and sealants (lf)

For some of these items, you will have to get material quotes from suppliers and add your own labor to install.

ESTIMATING DOORS

Perform a door count and verify the door schedule (if provided) with actual plans to locate each door. Depending on the types, doors will be quoted from different suppliers and contractors.

Use the following checklist (with measurement units).

Estimating Doors Checklist

- ☐ Hollow metal doors (ea)
- ☐ Wood doors (ea), including solid core, hollow core, prehung
- ☐ Bifold doors (ea)
- ☐ Wood pocket doors (ea)
- ☐ Mirrored bipass doors (ea)
- ☐ Patio doors (ea)
- ☐ Special doors (ea)
  - ☐ Overhead doors and grilles
  - ☐ Vertical lift doors
  - ☐ Automatic entrance/sliding doors
  - ☐ Traffic impact doors
  - ☐ Accordion folding doors
- ☐ Pressed steel frames (ea), including those for interior windows
- ☐ Finish hardware (ea)
- ☐ Miscellaneous glazing (sf)
- ☐ Installation (ea)

Download this form at www.**DeWALT**.com/guides

ESTIMATING STOREFRONT AND WINDOWS

For commercial/institutional projects, glass contractors will supply and install curtain walls, storefront doors, and glazing. Look for a door schedule listing all aluminum doors, their sizes, details for jambs, heads, and sills. Also check the plans and specs for approved manufacturers and types of glass.

In residential projects, vinyl windows are quoted supply only, so you must add installation costs. Look for a window schedule listing openings, window type and size, glazing, frame material and details, required accessories, and hardware.

Use the following checklist (with measurement units).

Estimating Storefront and Windows Checklist

- ❐ Storefront and curtain wall (sf)
- ❐ Aluminum glass doors (ea)
- ❐ Exterior aluminum or vinyl windows (ea or sf)
- ❐ Accessories (e.g., blueskins, screens, flashing, sills) (lf or sf)
- ❐ Sliding or patio doors (ea)
- ❐ Interior windows, glazing, and shower doors (ea or sf)
- ❐ Glazing (sf), including for balcony handrails, automatic doors
- ❐ Hardware for storefront doors (ea)
- ❐ Automatic operators (ea)
- ❐ Engineering (l/s)
- ❐ Installation (l/s)

Download this form at **www.DEWALT.com/guides**

ESTIMATING FINISHES

In commercial and institutional projects, architects will draft a detailed finish schedule, showing what finishes go on walls, floors, and ceilings, room by room. In residential jobs, there will be rough guidelines. For example, owners may want hardwood flooring for the dinning area, carpet in bedrooms, tile in bathrooms, and so forth.

If no room finish schedule is provided, then create one based on the owner's requirements. For each room listed, measure its area and perimeter. For example, in each rectangular room:

$$\text{Room Area} = \text{Room Length} \times \text{Room Width}$$
$$\text{Room Perimeter} = (\text{Room Length} + \text{Room Width}) \times 2$$

Then finish area can be calculated for each room:

- Floor Finish = Room Area
- Floor Base = Room Perimeter
- Wall Finish = Room Perimeter × Ceiling Height (if same finish on all four walls)
- Ceiling Finish = Room Area

When you add up each room, you have the total quantities for interior finishes.

Estimating Example

A 12 ft × 10 ft room with 8 ft ceiling:

$$\text{Room Area} = 12 \text{ ft} \times 10 \text{ ft} = 120 \text{ sf}$$
$$\text{Room Perimeter} = (12 + 10) \times 2 = 44 \text{ lf}$$
$$\text{Floor finish} = 120 \text{ sf}$$
$$\text{Floor base} = 44 \text{ lf}$$
$$\text{Wall finish} = 44 \times 8 = 352 \text{ sf}$$
$$\text{Ceiling finish} = 120 \text{ sf}$$

ESTIMATING DRYWALL

In residential construction, wood framing is done by a framer and the drywall contractor will come later to install gypsum boards on walls and ceilings. In commercial/institutional construction, a drywall subtrade will be responsible for light-gauge metal framing as well as gypsum boards. In either case, drywall quotes should be read carefully regarding what they include (e.g., insulation, vapor barrier, wood blocking, plywood sheathing).

Use the following checklist for drywall takeoff (with measurement units).

Estimating Drywall Checklist

- ☐ Stud and track metal partitions with bridging (sf or lf)
- ☐ Furring, wood or metal (lf)
- ☐ Gypsum wall board (sf)
- ☐ Drywall ceiling (sf) and bulkheads (lf)
- ☐ Densglass or plywood sheathing (sf)
- ☐ Wood blocking (lf or bf)
- ☐ Insulation and vapor barrier (sf)
- ☐ Corner bead (lf)
- ☐ Mud (lb, gallon, or bucket)
- ☐ Tape (lf or box)
- ☐ Screws and nails (ea or box)
- ☐ Scaffolding (sf)
- ☐ Cut and patch (l/s)
- ☐ Finishing (sf)

Download this form at www.**DEWALT**.com/guides

Estimating Math (Based on Common 4 × 8 Sheet)

Wall Board Area = Wall Length × Wall Height × Cover Sides × Board Layers
Number of Drywall Boards (ea) = Wall Board Area/32

Note that "wall board area" directly relates to the actual number of gypsum boards needed, which may differ from the area of wall to be covered.

Ready mix = 14 lb per 100 sf of wall board area (or 4.5 lb per sheet)
Tape = 48 lf per 100 sf of wall board area (or 15.4 lf per sheet)
Joint compounds = 9 lb per 100 sf of wall board area (or 2.9 lb per sheet)
Nails = 0.5 lb per 100 sf of wall board area (or 0.16 lb per sheet)
Screws = 125 ea per 100 sf of wall board area (or 40 ea per sheet)

Estimating Example

A 15 ft × 9 ft wall covered with double layers of drywall on both sides:

$$\text{Wall board area} = 15 \times 9 \times 2 \times 2 = 540 \text{ sf}$$
$$\text{Number of drywall boards} = 540/32 = 17 \text{ ea}$$
$$\text{Ready mix} = 14 \times 540/100 = 76 \text{ lb}$$
$$\text{Tape} = 48 \times 540/100 = 260 \text{ lf}$$
$$\text{Joint compounds} = 9 \times 540/100 = 49 \text{ lb}$$
$$\text{Nails} = 0.5 \times 540/100 = 2.7 \text{ lb}$$
$$\text{Screws} = 125 \times 540/100 = 675 \text{ ea}$$

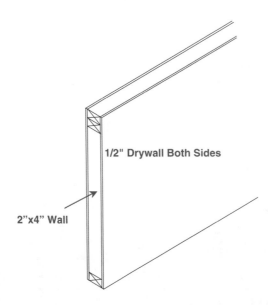

1/2" Drywall Both Sides

2"x4" Wall

ESTIMATING EIFS/STUCCO

Trade contractors often include EIFS/stucco in their drywall quote as a total package. For takeoff, it is customary to go over all sides of building exteriors to measure the finish area. Also check floor plans to include zigzag lines, because not all wall areas can be seen from exterior elevations.

In addition to wall areas, check other locations that may get EIFS/stucco finishes (e.g., ceilings, exterior soffits, canopy columns, under exterior balcony decks, roof and floor overhangs, roof parapets, retaining walls, dumpster walls, site features).

Use the following checklist for EIFS/stucco (with measurement units).

Estimating EIFS/Stucco Checklist

❏   Lath, gypsum, or metal (sy)

❏   Plaster (sy, cf, or bags)

❏   EIFS (sf)

❏   Trims, moldings, and shapes (lf or ea)

❏   GFRC columns (ea)

Download this form at **www.DEWALT.com/guides**

Carefully read specs on stucco to find out the number and thickness of coats, mixes to be used, and type of lath required.

ESTIMATING ACOUSTICAL CEILINGS

Sometimes drywall contractors will include acoustical ceilings in their price as a total package, but check their quotes to be sure.

Use the following checklist for acoustical ceilings (with measurement units).

Estimating Acoustical Ceilings Checklist

❏   Ceiling tile (sf or ea)

❏   Insulation (sf)

❏   Suspension system (sf)

  ❏   Main runner (lf)

  ❏   Cross runner (lf)

  ❏   Wall angle (lf)

  ❏   Struts (lf)

  ❏   Wire (ea)

  ❏   Seismic post (ea)

  ❏   Wall angle fasteners (ea)

  ❏   Pop and grid fasteners (ea)

❏   Installing special lighting fixtures (ea)

❏   Metal strips (lf)

Download this form at **www.DEWALT.com/guides**

ESTIMATING FLOORING

Use the following flooring takeoff checklist (with measurement units).

Estimating Flooring Checklist

- ❒ Carpet, including wall carpet (sf or sy)
- ❒ Carpet tile (sf or sy)
- ❒ Resilient flooring (sf or sy)
- ❒ Vinyl composition tile (sf or sy)
- ❒ Rubber flooring (sf or sy)
- ❒ Terrazzo (sf or sy)
- ❒ Ceramic tile (sf or sy) including wall tile, exterior deck tile, fireplace tile surrounds, and bull-nose trims
- ❒ Quarry or mosaic tile (sf or sy)
- ❒ Rubber tile (sf or sy)
- ❒ Marble or stone flooring (sf or sy)
- ❒ Laminate or hardwood flooring (sf or sy)
- ❒ Base (lf or y)
- ❒ Threshold (lf)
- ❒ Metal transition strips (lf)
- ❒ Epoxy grout (sf)
- ❒ Floor underlayment (sf)
- ❒ Flooring in elevators cabs (sf)
- ❒ Flooring on stairs (sf), treads (sf), and tactile warning strips (lf)

Download this form at **www.DeWALT.com/guides**

ESTIMATING PAINTING

Use the following painting checklist (with measurement units). Note that both exterior and interior of the building should be checked for painting scope and requirements (e.g., number of coats, brand of paint, inspection).

Estimating Painting Checklist

- ☐ Painting walls (sf), including exterior cladding and interior partitions
- ☐ Painting ceiling (sf), including covered and exposed areas
- ☐ Painting floor (sf)
- ☐ Painting columns and beams (sf or lf)
- ☐ Painting trims, moldings, cornices, and baseboards (lf)
- ☐ Painting stairs (sf or lf)
- ☐ Painting doors (ea or sf)
- ☐ Painting cabinets (ea or sf)
- ☐ Structural steel priming or painting (ton or sf)
- ☐ Vinyl wall covering (sf), including corner/bumper guards
- ☐ Painting exterior siding (sf)
- ☐ Painting downspouts and gutters (lf)
- ☐ Painting fences and railing (lf)
- ☐ Painting bollards (ea)
- ☐ Painting parking lines (lf) for exterior and interior parking spots
- ☐ Caulking (lf)
- ☐ Special waterproof coating to masonry or precast concrete walls (sf)

Download this form at **www.DeWALT.com/guides**

ESTIMATING SPECIALTIES

Specialties include a list of miscellaneous items to be installed near the end of the project. Before estimating, carefully go through both drawings and specs to see what is involved. Some items are too expensive to miss.

Following are common specialty items.

- Toilet partitions and urinals (ea)
- Washroom accessories, including hand-dryers (ea)
- Medicine cabinets (ea), if not by millwork contractor
- Washroom mirrors (ea), if not by glazing contractor
- Shower doors (ea), if not by glazing or plumbing contractor
- Chalkboards, tack boards, marker boards (ea)
- Mailboxes (ea)
- Flagpoles (ea)
- Bicycle racks (ea)

- Trash receptacles (ea)
- Awnings (lf, sf, or ea)
- Lockers and benches (ea)
- Shutters, grilles, louvers (ea)
- Metal corner guards (ea or lf)
- Fireplaces and stoves (ea)
- Signage (ea)

ESTIMATING EQUIPMENT

Equipment is not a division to be neglected. Some equipment is to be supplied by the owner, but requires you to install. Even if the package is totally by the owner, you may still be responsible for related work such as equipment receiving and storage, supporting brackets, plates and angles, mechanical and electrical connections, or wood blocking. Owner-supplied equipment frequently tends to be long-lead items to delay the project schedule, which will impact your overhead.

Common equipment items include the following:

- Residential appliances (ea)
  - Clothes washer and dryer
  - Dishwasher
  - Refrigerator
  - Range and hoods
- Parking control equipment (ea)
- Loading dock equipment (ea), including seals, bumpers, levelers, etc.
- Food service equipment (ea)
- Lab equipment (ea)
- Gym equipment (ea)
- Security and vault equipment (ea)
- Library equipment (ea)
- Medical equipment (ea)
- Playground equipment (ea)

ESTIMATING FURNISHINGS

Do not assume the owner will supply the furnishings. Check drawings and specs to make sure and ask for clarifications if necessary.

The following is a list of common furnishing items.

- Blinds and shades (ea, lf, or sf)
- Draperies and curtains (ea or lf)
- Church pews (ea)

- Bleachers (ea)
- Floor mats (ea)
- Casework (ea or lf)
- Display cases (ea or lf)
- Artwork (ea)
- Furniture (ea)
- Booths, tables, and chairs (ea)

In estimating furnishings, be sure to account for the following costs.

- Material supply and shipping
- Handling and assembly
- Installation and cleaning

ESTIMATING SPECIAL CONSTRUCTION

Usually special construction items are to be supplied and installed by a trade contractor as a complete package. Verify if all related costs such as excavation, unloading crane, and electrical wiring are included. If not, either ask the sub to add those to the quote, or allow additional money in your own budget.

Following are examples of special construction items.

- Swimming pool (l/s)
- Spa and sauna (l/s)
- Preengineered metal buildings (l/s)
- Greenhouses (l/s)
- Ice rinks (l/s)
- Storage tanks (l/s)
- Tennis courts (l/s)
- Vaults (l/s)
- Special purpose rooms (l/s)

Sometimes, fire sprinkler systems are listed under this division. Be sure to check if the mechanical contractor is including sprinklers in the quote.

ESTIMATING CONVEYING SYSTEMS

Look at drawings for locations and quantities of conveying systems, and then check specs to ensure you have the right one with correct type, capacity, manufacturer's name.

Common conveying systems are as follows:

- Passenger elevators (ea), hydraulic or electrical
- Freight elevators (ea)
- Pneumatic tubing system (ea)
- Chutes, including laundry, linen, or waste (ea)

- Dumbwaiters (ea)
- Handicap lifts (ea)
- Escalators (ea)
- Moving walks (ea)

Consider all related costs such as elevator door, cab floor finish, hoist or divider beams, and electrical wiring. Include additional cost in the estimate if information is not available.

ESTIMATING PLUMBING

Quantity Takeoff Procedures

Take the following steps

1. Count each type of fixture and equipment for each floor or section of the building. Total them separately.

2. Start with the largest size in the pipe line, follow along the main, check off and note the number of valves and tees for that size, as well as the size and number of risers.

3. Mark the point where the size of the line changes.

4. Return back down the same line, check and take off the number of fittings.

5. Start again from the beginning, measure the footage of pipe and round off to the nearest 10 ft. Then mark the line with a colored pencil to indicate you have completed the takeoff.

6. Follow the same procedure for each size of pipe; continue to the end of the main or riser. Then work from the end of the main, return back down the line and take off the branches: valves, fittings, and pipes to the last outlet.

7. Recheck the length of the risers for distance between floors from architectural or structural plans.

8. Carefully check the drawing for unmarked pipe lines to ensure that nothing has been missed.

9. Total all quantities for each different system independently.

On small jobs (such as a house), plumbers will likely perform all the work inside and outside of the building. On large jobs, however, the plumbing or sprinkler contractor is only responsible for the work within a few feet of the building footprint. Anything beyond that boundary is picked up by the site servicing contractor.

The next few pages provide detailed quantity takeoff checklists.

Estimating Plumbing Checklist

Coordination with Exterior Site Utilities

- ❑ Building sewer connection (l/s)
- ❑ Building water connection (l/s)
- ❑ Roof drain connection (l/s)
- ❑ Irrigation system coordination (l/s)
- ❑ Pipe demolition (lf)
- ❑ Grease/oil/lint interceptor (ea)
- ❑ Sump pump (ea)
- ❑ Septic tank (ea)
- ❑ Water meter (ea)
- ❑ Domestic water backflow preventer (ea)
- ❑ Foundation drain tile (lf) including filter cloth and drain rock

Plumbing Fixtures

- ❑ Drinking fountains (ea)
- ❑ Wash fountains (ea)
- ❑ Bubblers (ea)
- ❑ Water closets (ea)
- ❑ Bidets (ea)
- ❑ Urinal (ea)
- ❑ Bath tubs (ea)
- ❑ Showers (ea)
- ❑ Laundry tubs (ea)
- ❑ Lavatories (ea)
- ❑ Sinks (ea)
- ❑ Garbage can wash (ea)
- ❑ Garbage disposer (ea)
- ❑ Connections for dishwashers (ea)
- ❑ Connections for clothes washers (ea)
- ❑ Connections for ice maker (ea)
- ❑ Wall hydrants (ea)
- ❑ Hose bibs (ea)
- ❑ Fixture trims and carrier(ea)
- ❑ Fixture cleaning (l/s)

(continues)

❏ Testing and adjustment (l/s)

Plumbing Equipment

❏ Sewage pump (ea)
❏ Sump pump (ea)
❏ Circulating pump (ea)
❏ Pressure booster pump (ea)
❏ Backflow preventer (ea)
❏ Water heater (ea)
❏ Storage tank (ea)
❏ Expansion tank (ea)
❏ Shutoff valve (ea)
❏ Pressure reducing valve (ea)
❏ Thermostatic mixing valve (ea)
❏ Siphon breaker (ea)
❏ Air compressor (ea)
❏ Sand/oil interceptor (ea)
❏ Grease trap (ea)
❏ Water softener (ea)
❏ Testing and adjustment (l/s)

Interior Plumbing Piping

❏ Excavation and backfill (cy)
❏ Drainage, waste, and vent pipe (lf)
❏ Indirect waste pipe (lf)
❏ Rainwater leader (lf)
❏ Hot water pipe (lf)
❏ Cold water pipe (lf)
❏ Sanitary waste pipe (lf)
❏ Sanitary vent pipe (lf)
❏ Sleeve (lf)
❏ Fitting (ea)
❏ Valve and control device (ea)
❏ Trench drain (lf)
❏ Floor drain (ea)

(continues)

❏ Roof drain (ea)

❏ Deck drain (ea)

❏ Planter drain (ea)

❏ Floor sink (ea)

❏ Clean-out (ea)

❏ Insulation (lf or sf)

❏ Hangers and support (l/s)

Special Piping Systems

❏ Natural gas piping (l/s)

❏ Chilled water piping (l/s)

❏ High temperature water piping (l/s)

❏ Tempered water piping (l/s)

❏ Distilled water piping (l/s)

❏ Carbon dioxide piping (l/s)

❏ Compressed air piping (l/s)

❏ Medical oxygen piping (l/s)

❏ Medical vacuum piping (l/s)

❏ Medical air piping (l/s)

❏ Nitrous oxide piping (l/s)

❏ Nitrogen piping (l/s)

❏ Process piping (l/s)

❏ Chemical piping (l/s)

❏ Laboratory waste and vent piping (l/s)

❏ Kitchen equipment piping (l/s)

❏ Liquid soap piping (l/s)

Download this form at **www.DeWALT.com/guides**

If a project requires any special piping systems, then ask the plumber to include these in the bid, or obtain an additional quote from another source.

Estimating Special Piping Systems Checklist

Miscellaneous

- ❒ Roof flashing (lb)
- ❒ Caulking compound (gallon)
- ❒ Caulking lead (lb)
- ❒ Oakum (lb)
- ❒ Solder/flux (lb)
- ❒ Presto gas tank (l/s)
- ❒ Concrete insert (l/s)
- ❒ Pipe support hanger (l/s)
- ❒ Vibration isolation (l/s)
- ❒ Insulation for piping and equipment (l/s)
- ❒ Wood backing (l/s)
- ❒ Washroom accessories (ea)
- ❒ Access panels (ea)
- ❒ Valve tags (ea)
- ❒ Covers and frames (ea)
- ❒ Rigging (l/s)
- ❒ Pipe painting (l/s or sf)
- ❒ Excavation and backfill (cy)
- ❒ Concrete pad (ea or sf)
- ❒ Small tools (l/s)

Download this form at **www.DEWALT.com/guides**

ESTIMATING FIRE SPRINKLERS

Fire sprinkler systems are usually installed by a separate contractor, not the plumber. Coordinate the connection with exterior site serving work, and consider costs for fire line backflow preventers, fire department connection, fire hydrants, and other associated items.

Use the following takeoff checklist (with measurement units).

Estimating Fire Sprinklers Checklist

- ❏ Backflow preventer (ea)
- ❏ Fire pump (ea)
- ❏ Siamese connection (ea)
- ❏ Sprinkler valve station (ea)
- ❏ Fire hose cabinet and racks (ea)
- ❏ Sprinkler standpipe (lf)
- ❏ Standpipe pressure pump (ea)
- ❏ Jockey pump (ea)
- ❏ Sprinkler head (ea)
- ❏ Fittings (ea)
- ❏ Valves and control devices (ea)
- ❏ Stairway fire department valve (ea)
- ❏ Automatic alarm calve (ea)
- ❏ Fire extinguisher (ea)
- ❏ Hangers and support (l/s)
- ❏ Test and balance (l/s)

Download this form at **www.DEWALT.com/guides**

ESTIMATING HVAC

Some mechanical contractors will quote HVAC, plumbing, and/or fire sprinklers as a total package. Carefully read their quotes to see what they have included. The next few pages provide detailed HVAC takeoff checklists.

Estimating HVAC Checklist

Equipment

- ❏ Boilers, burners, furnaces (ea)
- ❏ Unit heaters, convectors (ea)
- ❏ Compressor or chiller units (ea)
- ❏ Condensers (ea)
- ❏ Receivers (ea)
- ❏ Cooling towers (ea)
- ❏ Chimneys (ea)
- ❏ Heat exchangers (ea)
- ❏ Air handling units (ea)
- ❏ Exhaust fans (ea)
- ❏ Ventilators (ea)
- ❏ Expansion tanks (ea)
- ❏ Storage tanks (ea)
- ❏ Heat pumps (ea)
- ❏ Condensate pumps (ea)
- ❏ Dust collector (ea)
- ❏ Fume hoods (ea)

Piping

- ❏ Chilled water piping (lf)
- ❏ Hot water piping (lf)
- ❏ Condenser water piping (lf)
- ❏ Refrigerant lines piping (lf)
- ❏ Steam piping (lf)
- ❏ Condensate piping (lf)
- ❏ Oil piping (lf)
- ❏ Gas piping (lf)
- ❏ Pipe insulation (lf)
- ❏ Fittings and valves (ea)
- ❏ Pipe painting (l/s or sf)

(continues)

Ductwork

- ❏ Supply ducts (lf, sf, lb)
- ❏ Return ducts (lf, sf, lb)
- ❏ Ductwork insulation (sf)
- ❏ Louvers, diffusers, registers, dampers, and grilles (ea)
- ❏ Fittings (ea)
- ❏ Valves (ea)
- ❏ Filters (ea)

Miscellaneous

- ❏ Packaged control (ea)
- ❏ Starters (ea)
- ❏ Motors (ea)
- ❏ Thermostat (ea)
- ❏ Humidistat (ea)
- ❏ Air purification (l/s)
- ❏ Humidification and dehumidification (l/s)
- ❏ Test and balance (l/s)
- ❏ Refrigerant (lb)
- ❏ Vibration isolation (l/s)
- ❏ Access panels (ea)
- ❏ Valve tags (ea)
- ❏ Covers and frames (ea)
- ❏ Rigging (l/s)
- ❏ Equipment room construction (l/s)
- ❏ Concrete pad (ea or sf)
- ❏ Excavation and backfill (l/s)
- ❏ Electrical wiring (l/s)

Download this form at **www.DeWALT.com/guides**

ESTIMATING ELECTRICAL
Electrical Takeoff Sequence

- Lighting fixtures and lamps
- Wiring devices
- Conduits, connectors, and wire
- Boxes and covers
- Service and distribution (one-line diagram)
- Special systems
- Site electrical
- Equipment hookup
- Miscellaneous material and installation

The next few pages provide detailed electrical takeoff checklists.

Estimating Electrical Checklist

Switchgear

- ☐ Main switchboards (ea)
- ☐ Distribution panel boards (ea)
- ☐ Load centers (ea)
- ☐ Motor control centers (ea)
- ☐ Motor starters (ea)
- ☐ Power panels (ea)
- ☐ Lighting panels (ea)
- ☐ Transformers (ea)
- ☐ Terminal cabinets (ea)
- ☐ Feeder cables (lf)
- ☐ Circuit breakers (ea)
- ☐ Fuses (ea)
- ☐ Meters and meter centers (ea)
- ☐ Capacitors (ea)
- ☐ Safety switches (ea)
- ☐ Manual starters (ea)
- ☐ Lighting contactors (ea)
- ☐ Bus ducts (lf)
- ☐ Time clocks (ea)
- ☐ Push-button stations (ea)

(continues)

Lighting Fixtures

- ❏ Fixtures (ea)
- ❏ Lamps, including incandescent, fluorescent, mercury vapor (ea)
- ❏ Plaster frames (ea)
- ❏ Remote ballasts (ea)
- ❏ Hangers (ea)
- ❏ Stems (ea)
- ❏ Spacers (ea)
- ❏ Couplings (ea)
- ❏ Tandem units (ea)
- ❏ Suspension systems (l/s)
- ❏ Floodlight poles (ea)
- ❏ Brackets (ea)
- ❏ Valance and cove lighting (ea)
- ❏ Special lenses (e.g., plastic or glass lenses, parabolic louvers)
- ❏ Egg crate and reflectors (ea)
- ❏ Signs (ea)
- ❏ Security lighting (ea)

Wiring Devices

- ❏ Switches, dimmers, sensors (ea)
- ❏ Receptacle, all types (ea)
- ❏ Finish plates, all types (ea)
- ❏ Special outlets (ea)
- ❏ Time switches and photo cells (ea)

Conduit and Wires

- ❏ Conduits, all types including rigid, EMT, flexible, PVC (lf)
- ❏ Underfloor and wall ducts (lf)
- ❏ Cable tray (lf)
- ❏ Conduit fittings, including elbows, locknuts, bushings, straps, clamps, hangers, condulets, expansion fittings, nipples (ea)
- ❏ Enclosures and cabinets (ea)
- ❏ Terminals and lugs (ea)
- ❏ Wire and cable, all types (lf)
- ❏ Grounding (l/s)
- ❏ Outlet boxes and plaster rings (ea)

(continues)

Special Systems

- ❑ Site lighting (l/s)
- ❑ Fire alarm and signaling (l/s)
- ❑ Telephone, intercom, cable and data (l/s)
- ❑ Audio and video (l/s)
- ❑ Security and CCTV (l/s)
- ❑ Underfloor duct (l/s)
- ❑ Clocks and clock systems (l/s)
- ❑ Hospital including nurse call, doctor's register, others (l/s)
- ❑ Lightning protection (l/s)
- ❑ Electric heat, snow melting, and heat tracing (l/s)
- ❑ Emergency power and UPS (l/s)
- ❑ Energy management (l/s)
- ❑ Substations (l/s)
- ❑ Overhead transmission (l/s)

Miscellaneous

- ❑ Equipment hookup (l/s)
- ❑ Trenching, excavation, concrete, core drilling, light bases (l/s)
- ❑ Temporary power and lighting (l/s)
- ❑ Demolition of existing electrical system (l/s)
- ❑ Mechanical control wiring and starters (l/s)
- ❑ Forklifts, cranes, and special hoisting (l/s)
- ❑ Equipment and tools (l/s)
- ❑ Access panels (ea)
- ❑ Cutting and patching (l/s)
- ❑ Roof penetration (ea)
- ❑ Fixture support (ea)
- ❑ Panel backing (l/s)
- ❑ Scaffolding (l/s)
- ❑ Testing (l/s)
- ❑ Painting (l/s)
- ❑ Dust protection (l/s)
- ❑ Firestopping and smoke seal (l/s)
- ❑ Sleeves (lf)
- ❑ Primary transformer (ea)
- ❑ Primary raceway (lf)
- ❑ Primary cable (lf)

CHAPTER 3

PRICING

Pricing is the process of converting the quantity takeoff into dollar values. It is not about making wild guesses, but identifying all cost items to determine the most accurate price, regardless of the project's size.

An important principle to keep in mind is that you are pricing labor and material according to the time when the work is expected to be done, not when the job is being estimated. Some work is done several months after the bid is submitted. It is impossible to predict what the prices will be in three to six months. Therefore, if you expect a price escalation or labor shortage in the future, it is better to make allowances now, or obtain a written price guarantee from your suppliers or subcontractors.

In this chapter, the following topics will be covered.

- Estimating material costs
- Evaluating material suppliers
- Estimating labor costs
- Man-hour estimates and adjustments
- Estimating equipment costs
- Combined pricing summary worksheet

PRICING PROCEDURES

The following are general guidelines for pricing.

1. Finish takeoff first. Summarize the material quantities.
2. Combine quantities for the same items and allow for reasonable waste.
3. Transfer only the total quantities to your pricing sheets.
4. Pricing materials based on the quotes you received from the suppliers.
5. Pricing labor based on the adjusted productivity (man-hour) information.
6. Add quotes from subcontractors or suppliers if necessary.
7. Add indirect costs (e.g., overhead, bond, insurance, permits).
8. Allow costs for items that are not shown on drawings but required.
9. Add owner's cash allowance for this portion of work.
10. Allow contingencies due to problems in design and field construction.
11. Add profits to get a total price.

If you are a general contractor self-performing the work, then your costs for this portion of work (e.g., form-work or framing) should not be net. You should include some profit and overhead in the price, just like most subcontractors do.

ESTIMATING MATERIAL COSTS

The basic formula for estimating materials costs is:

$$\text{Material Price} = \text{Quantity} \times \text{Material Unit Price}$$

Do Your Own Takeoff

For self-performed work (or when buying materials and having someone else do the installation), request quotes from at least three material suppliers. It is essential to have your own quantity takeoff. Suppliers may have their own estimating services, but because they do not install, their estimate may omit important items that you have to pay later as extras.

Get Quotes

As mentioned, request quotes from at least three suppliers. Attach a copy of your takeoff. Specify as much additional information as possible (e.g., concrete strengths, lumber grades, product type, model number and make). Sometimes it is necessary to furnish the specs (or even drawings) for your suppliers.

Evaluate Quotes

When quotes come in, read them carefully to verify the following factors.

- Unit price (some conversions might be necessary)
- Delivery charge (ideally it should say FOB jobsite)
- Sales taxes (including federal, state, city, and county taxes)
- Minimum order quantity
- Expected price escalation (e.g., increases for next year)
- Storage costs or standby charges
- Discount rates (apply it with caution)
- What is included (e.g., Are framing connectors included in lumber supply package?)

Certain materials are not available to some suppliers and they may propose an alternate; but it is hard to determine at the time of the bid whether an alternate is equal to the specified item. You may check with the architect or owner to see if the alternate is acceptable, but some bargain items may require too much labor to install or too much effort to get approved.

If one supplier's estimate is too low, then ask him to make corrections to all bidding contractors. To be ethical in your business practices, do not mention any details regarding other quotes to this supplier, as every quote is supposed to be confidential.

MATERIAL PRICING WORKSHEET

ABC Contracting
1 Main Street
Anytown, USA 00000
(555) 555-1234

Job Name: ABC School   Estimate No. 901   Estimator: AD
Date: Jan 1st 20XX   Worksheet Page Number: P1

Item	Quantity	Unit	Unit Price	Extension
A	6	EA	$130.00	$ 780.00
B	2	EA	$ 40.00	$ 80.00
C	4	EA	$ 50.00	$ 200.00
D	15	EA	$135.00	$ 2,025.00
E	3	EA	$ 50.00	$ 150.00
F	1	EA	$ 90.00	$ 90.00
G	3	EA	$100.00	$ 300.00
H	3	EA	$ 25.00	$ 75.00
I	2	EA	$ 35.00	$ 70.00
J	1	EA	$ 40.00	$ 40.00
K	1	EA	$ 69.00	$ 69.00
L	1	EA	$ 75.00	$ 75.00
M	1	EA	$180.00	$ 180.00
N	2	EA	$ 90.00	$ 180.00
O	3	EA	$ 45.00	$ 135.00
P	2	EA	$ 40.00	$ 80.00
Q	6	EA	$ 92.00	$ 552.00
Subtotal				**$5,081.00**
Sales Tax	6%			$ 304.86
Freight	1%			$ 50.81
Total Material Costs				**$5,436.67**

Download this form at **www.DEWALT.com/guides**

EVALUATING MATERIALS SUPPLIERS

Item	Quantity	ACME Ltd.		ABC Inc.		XYZ Corp.	
		Unit Price	Extension	Unit Price	Extension	Unit Price	Extension
A	6	$ 125.00	$ 750.00	$ 130.00	$ 780.00	$ 130.00	$ 780.00
B	2	$ 30.25	$ 60.50	$ 35.25	$ 70.50	$ 36.25	$ 72.50
C	4	$ 40.50	$ 162.00	$ 45.50	$ 182.00	$ 46.50	$ 186.00
D	15	$ 121.00	$ 1,815.00	$ 126.00	$ 1,890.00	$ 127.00	$ 1,905.00
E	3	$ 40.00	$ 120.00	$ 45.00	$ 135.00	$ 46.00	$ 138.00
F	1	$ 79.00	$ 79.00	$ 84.00	$ 84.00	$ 82.00	$ 82.00
G	3	$ 94.85	$ 284.55	$ 92.85	$ 278.55	$ 90.85	$ 272.55
H	3	$ 19.95	$ 59.85	$ 17.95	$ 53.85	$ 15.95	$ 47.85
I	2	$ 31.00	$ 62.00	$ 29.00	$ 58.00	$ 27.00	$ 54.00
J	1	$ 34.25	$ 34.25	$ 32.25	$ 32.25	$ 32.75	$ 32.75
K	1	$ 64.00	$ 64.00	$ 62.00	$ 62.00	$ 62.50	$ 62.50
L	1	$ 71.00	$ 71.00	$ 69.00	$ 69.00	$ 69.50	$ 69.50
M	1	$ 175.00	$ 175.00	$ 173.00	$ 173.00	$ 173.50	$ 173.50
N	2	$ 83.00	$ 166.00	$ 84.00	$ 168.00	$ 82.25	$ 164.50
O	3	$ 35.00	$ 105.00	$ 36.00	$ 108.00	$ 35.00	$ 105.00
P	2	$ 38.00	$ 76.00	$ 39.00	$ 78.00	$ 38.00	$ 76.00
Q	6	$ 90.00	$ 540.00	$ 91.00	$ 546.00	$ 90.00	$ 540.00
Subtotal			**$ 4,624.15**		**$ 4,768.15**		**$ 4,761.65**
Sales Tax	6%		$ 277.45		$ 286.09		$ 285.70
Freight	1%		$ 46.24		Included		Included
Total			**$ 4,947.84**		**$ 5,054.24**		**$ 5,047.35**

ESTIMATING LABOR COSTS

The basic formula for estimating labor costs is:

$$\text{Total Man-hours} = \text{Quantity} \times \text{Man-hour/Unit}$$
$$\text{Labor Hourly Rate} = \text{Basic Wage} \times (1 + \text{Labor Burden Rate})$$
$$\text{Total Labor Price} = \text{Total Man-hours} \times \text{Average Crew Hourly Rate}$$

It is more difficult to control labor costs than material costs, because labor is subject to too many variables. Generally the process can be done in five steps.

1. *Determine man-hours per unit:* Man-hour per unit is how long it takes one person to do one unit of work. This can be obtained by tracking historical productivity information from job to job. For example, a carpenter and a helper spent an 8-hour day to install 10 wood doors, so total man-hours are 2 people × 8 hours = 16 man-hours. Thus, one wood door will take 16 hours/10 doors = 1.6 man-hours per door.

2. *Estimate total man-hours:* Adjust historical man-hour numbers for the job at hand. Consider factors that can influence productivity, such as job size, overtime, crew, delays, height, site congestion, and weather. For example, the current job has 800 wood doors, but your carpenter recently quit and new installer is too green. So you decide it will now take 3.2 hours to install one wood door, instead of 1.6 hours originally calculated. So the total man-hours are 800 doors $\times$ 3.2 hours/door = 2,560 man-hours.

3. *Figure labor burden rate:* Labor burdens are all the extras involved, such as fringe benefits in addition to the basic wage. Talk with your bookkeeper to obtain the following information.
   A. Total basic wages you paid to your crew last year (excluding any burdens)
   B. Total labor burdens you paid last year, including:
   - Taxable fringe benefits (e.g., vacation pay)
   - Tax-deferred pension or profit sharing plans
   - Medical insurance (health, dental, life, and disability)
   - Worker's compensation insurance
   - General liability insurance
   - Living allowances and cash compensation
   - Social security and Medicare taxes (FICA)
   - Federal unemployment tax (FUTA)
   - State unemployment tax (SUTA)
   - Union dues

   Suppose the total wages you paid last year were $150,000, while labor burden was an additional $36,000. The labor burden rate is then $36,000/$150,000 = 24%.

4. *Calculate crew rate:* First, for each member of the crew, take the basic wage and add labor burden to determine labor hourly rate. The crew rate will be the average of the team. For example, your crew for door installation is made of one carpenter and one helper. The carpenter is earning $30 per hour, so the adjusted wage with labor burden is $30 $\times$ (1 + 24%) = $37.20 per hour. The helper is earning $20 per hour, so the adjusted wage with labor burden is $20 $\times$ (1 + 24%) = $24.80 per hour. The overall crew rate will be the average of the two: ($37.20 + $24.80)/2 = $31 per hour.

5. *Subtotal labor costs:* To install 800 wood doors, you figured 2,560 man-hours at $31 per hour, so total labor cost is 2,560 $\times$ $31 = $79,360.

You may be asking yourself, "Why can't I skip all these troubles and randomly pick a labor unit rate? For example, for 800 wood doors, just say it takes $100 to install for each door, then the labor cost is simply 800 $\times$ $100 = $80,000. Isn't that close enough?"

The problem with this shortcut method is that $100 is a guess and not related to the project specific situations. What if labor productivity changes? What if wages increase? Every job is unique, and your pricing should reflect that fact.

LABOR PRICING WORKSHEET

ABC Contracting
1 Main Street
Anytown, USA 00000
(555) 555-1234

Job Name: ABC School
Date: Jan 1st 20XX

Estimate No. 901
Worksheet Page Number: P2

Estimator: AD

Item	Quantity	Unit	Man-hour	Extension
A	6	EA	0.60	3.60
B	2	EA	0.40	0.80
C	4	EA	1.00	4.00
D	15	EA	0.50	7.50
E	3	EA	1.10	3.30
F	1	EA	1.20	1.20
G	3	EA	0.50	1.50
H	3	EA	0.60	1.80
I	2	EA	0.60	1.20
J	1	EA	1.00	1.00
K	1	EA	1.20	1.20
L	1	EA	1.10	1.10
M	1	EA	1.50	1.50
N	2	EA	2.00	4.00
O	3	EA	0.40	1.20
P	2	EA	0.30	0.60
Q	6	EA	0.25	1.50
Total Man-hours				37.00
Crew Hourly Rate				$50.00
Labor Burden				Included
Labor Cost Subtotal				**$1,850.00**

Download this form at **www.DeWALT.com/guides**

ESTIMATING MAN-HOURS

Some contractors may apply a flat labor unit price rate (i.e., how much it costs to install each item) to their quantities, but unit prices are only as good as the experience behind them, especially when they fluctuate quite regularly. What works on one job may not work on another. Even contractors who have good instincts cannot successfully compensate for the differences from job to job.

A better way of estimating is to use man-hours, which, as mentioned, reflects what a person can do within an hour of work. This information tends to remain relatively stable from job to job. For example, from previous jobs you did, it seems to take 8 minutes for a three-person crew (i.e., 0.4 hour) to handle and install one sheet of wall panels. The crew is paid $50 per hour on average including labor burdens. Therefore, labor to install two wall panels is 2 ea × 0.4 man-hour × $50/man-hour = $40.

Here's how to determine the man-hour per unit:

Total Man-hours = Number of Working Crew Members × Hours Crew Worked
Unit Man-hour for the Item = Total Man-hours/Crew Output

Estimating Example

An electrician and a helper spent one 8-hour day to install 200 ft of conduits.

Total man-hours are 2 people × 8 hours = 16 hours.
One linear foot of such conduit will take 16 hours/200 ft = 0.08 hour/ft

MAN-HOUR TABLES

The next few pages provide sample man-hour tables. They are based on an average man working under normal conditions: new construction with fair productivity, standard materials and straight-forward installation, appropriate tools, and good coordination with other trades. You will need to make adjustments based on specific job conditions.

The man-hours listed in these tables include:

- Unloading, storing, and getting raw materials
- Getting and returning tools/equipment
- Normal time lost due to work breaks
- Reading drawings and discussing the work to be performed
- Normal handling, measuring, cutting, and fitting
- Field measurement and layout
- Test and balance of system
- Supervision time (foremen and superintendents, etc.)
- Setting up and tearing down of lifts and ladders
- Regular cleanup of construction debris
- Infrequent correction or repairs required because of faulty installation

Please note neither the author nor the publisher guarantees the accuracy of man-hour information in these tables. Use your judgment and caution.

Site Work

Item	Unit	Labor Hours
Site Demolition		
Clear & grub	acre	12 to 64
Tree removal	ea	2.5 to 7.0
Entire building demolition	sf	0.1 to 0.2
Paving removal	sy	0.1 to 0.5
Debris removal	cy	0.6 to 0.8
Building Interior Demo		
Demolishing footings	lf	0.1 to 0.3
Demolishing foundation wall	sf	0.1 to 0.2
Demolishing concrete slab	sf	0.05 to 0.10
Removing floor finish	sy	0.1 to 0.3
Demolishing wood framed wall	sf	0.05 to 0.10
Demolishing masonry block walls	sf	0.1 to 0.2
Demolishing ceiling	sf	0.02 to 0.05
Removing doors and frames	ea	0.5 to 1.5
Removing window and frame	ea	0.3 to 1.0
Demolishing roof structure and finish	sq	2.0 to 5.0
Earthwork		
Excavation by hand	cy	1.1 to 2.3
Excavate trench by hand	lf	0.5 to 1.3
Moving excavated materials by hand	cy	1.0 to 2.0
Backfilling by hand	cy	0.4 to 0.7
Spreading soil piled on site by hand	cy	0.2 to 0.4
Paving		
Concrete curbs	lf	0.2 to 0.3
Driveway paving	sy	0.2 to 0.3
Asphalt paving	sy	0.1 to 0.2
Landscape pavers	sy	1.3 to 2.1

Concrete

Item	Unit	Labor Hours
Formwork		
Footers	SFCA	0.1 to 0.2
Pads	SFCA	0.1 to 0.2
Foundation walls	SFCA	0.2 to 0.5
Slabs	SFCA	0.8 to 1.2
Beams	SFCA	0.1 to 0.5
Columns	SFCA	0.1 to 0.2
Placing Concrete		
Footers	cy	0.4 to 0.8
Pads	cy	0.6 to 1.3
Foundation walls	cy	0.5 to 0.9
Grade beams	cy	0.3 to 0.7
Slab on grade	cy	0.4 to 0.6
Elevated slab	cy	0.4 to 0.8
Beams	cy	1.0 to 1.8
Columns	cy	0.4 to 1.3
Placing Reinforcing Steel		
Footers	ton	9 to 16
Pads	ton	9 to 16
Foundation walls	ton	8 to 11
Slabs	ton	13 to 15
Beams	ton	12 to 21
Columns	ton	13 to 22
Finishing Concrete		
Slab on grade	sq	1.0 to 3.5
Walls	sq	3.0 to 3.5
Stairs	sq	3.0 to 4.0
Curbs	sq	2.5 to 3.0
Placing anchor bolts	ea	0.05 to 0.10
Placing embeds	ea	0.25 to 0.50

Masonry

Item	Unit	Labor Hours	
		Mason	*Laborer*
Foundation CMU Wall			
4" thick	sq	3.5 to 5.5	3.0 to 5.5
6" thick	sq	4.0 to 6.0	3.5 to 6
8" thick	sq	5 to 7	4.5 to 7.5
10" thick	sq	6 to 9.5	7 to 11
12" thick	sq	7 to 10	8 to 12
Exterior CMU Wall (up to 4' high)			
4" thick	sq	3.5 to 5.5	3.5 to 6.0
6" thick	sq	4.0 to 6.0	4.0 to 6.5
8" thick	sq	4.5 to 6.0	5.0 to 7.5
10" thick	sq	6.0 to 9.0	7.0 to 10.5
12" thick	sq	7.0 to 10.0	8.0 to 11.5
Exterior CMU Wall (4' to 8' high)			
4" thick	sq	3.5 to 6.0	4.5 to 7.5
6" thick	sq	4.0 to 6.5	4.5 to 7.0
8" thick	sq	4.5 to 6.5	6.0 to 9.0
10" thick	sq	6.5 to 10.0	7.5 to 12.0
12" thick	sq	7.5 to 10.0	8.5 to 12.0
Exterior CMU Wall (8' high and above)			
4" thick	sq	4.5 to 8.0	6.0 to 9.5
6" thick	sq	5.0 to 9.0	7.0 to 10.0
8" thick	sq	5.0 to 7.0	7.0 to 10.0
10" thick	sq	7.0 to 10.5	7.5 to 12.0
12" thick	sq	7.5 to 10.0	8.5 to 12.0
Interior Block Wall			
4" thick	sq	3.0 to 6.0	3.5 to 7.0
6" thick	sq	3.5 to 6.5	4.5 to 7.5
8" thick	sq	4.5 to 6.0	5.0 to 7.5
Face Brick (Standard/Modular)			
Common bond	sq	10.0 to 15.0	11.0 to 15.5
Running bond	sq	8.0 to 12.0	9.0 to 13.0
Stack bond	sq	12.0 to 18.0	10.0 to 15.0
Flemish bond	sq	11.0 to 16.0	12.0 to 16.0
English bond	sq	11.0 to 16.0	12.0 to 16.0

Item	Unit	Labor Hours	
Face Brick (Oversize)			
Common bond	sq	12.5 to 19.0	14.0 to 19.5
Running bond	sq	12.0 to 15.0	11.5 to 16.5
Stack bond	sq	15.0 to 22.5	12.5 to 19.0
Flemish bond	sq	14.0 to 20.0	15.0 to 20.0
English bond	sq	14.0 to 20.0	15.0 to 20.0

Steel

Item	Unit	Labor Hours
Structural Metal Framing		
Residential	ton	7.8 to 9.3
Commercial/Institutional	ton	7.8 to 9.3
Industrial	ton	6.2 to 7.6
Steel Joists		
K Series	ton	4.7 to 9.8
CS Series	ton	5.3 to 8.9
LH, DLH & SLH Series	ton	5.0 to 7.3
Joist Girders	ton	5.3 to 7.3
Steel Decking		
22 to 14 Gage	sq	1.8 to 3.5

Carpentry and Framing

Item	Unit	Labor Hours
Sills	1,000 bf	19 to 40
Sills/ledgers	100 lf	2 to 5
Girders	1,000 bf	8 to 33
Girders/beams, to 20' long	ea	1.0 to 1.5
Joists, built-up	1,000 bf	16 to 24
Joists, pre-engineered	1,000 lf	15 to 19
Wall Framing	1,000 bf	18 to 30
Wall Framing, 16" o.c., to 10'high	100 sf	13 to 16
Rafters	1,000 bf	19 to 35
Rafters	100 lf	2 to 4
Trusses, to 40' span	ea	0.6 to 1.0
Decking	1,000 bf	13 to 20

Item	Unit	Labor Hours
Subflooring	1,000 bf	12 to 20
Timber Framing	1,000 bf	10 to 24
Wall Sheathing	1,000 sf	12 to 20
Trim	100 lf	3 to 5
Blocking	100 lf	3 to 4
Fascia	100 lf	4 to 6
Soffits	sq	4 to 6
Framing for door/window opening	ea	1.5 to 3.5
Install steel doors	ea	1.5 to 2.5
Install wood doors	ea	1.0 to 2.5
Install door frames	ea	1.0 to 2.0
Install hardware set	ea	1.0 to 2.0
Install windows	ea	2.0 to 4.0

Roofing and Finishes

Item	Unit	Labor Hours
Roofing		
Asphalt shingle	sq	2.0 to 3.4
Wood shingle	sq	2.5 to 4.5
Clay tile	sq	4.5 to 6.0
Concrete tile	sq	5.5 to 6.0
Slate	sq	4.8 to 11.5
Metal	sq	3.0 to 4.5
Built-Up		
2 Ply	sq	1.3 to 2.3
3 Ply	sq	1.5 to 2.3
4 Ply	sq	1.7 to 2.6
5 Ply	sq	1.8 to 2.8
Aggregate surface	sq	0.3 to 0.5
Siding	sq	2.0 to 4.0
Building paper	sq	0.2 to 0.4
Insulation	sq	1.0 to 1.6
Drywall		
Gypsum board on wall	sq	2.0 to 4.0
Gypsum board on ceiling	sq	3.0 to 4.5
Metal framing	sq	1.5 to 3.5

Item	Unit	Labor Hours
Flooring		
Wood	sq	3.5 to 8.0
Sheet Vinyl	sq	3.0 to 4.0
Carpet	sq	2.5 to 3.5
Ceramic	sy	2.0 to 3.5
Terrazzo	sy	2.0 to 5.0
Acoustical ceiling, suspended	sq	2.0 to 3.5
Stucco	sq	5.5 to 8.0
Painting	sq	0.5 to 1.5
Wallpaper	sq	1.0 to 3.0

ADJUSTING MAN-HOURS

A common mistake is to forget that man-hours vary with installation conditions. Installation times can change drastically from job to job, from crew to crew, and even for the same crew from day to day. Use the following worksheet to adjust the standard man-hours.

ABC Contracting
1 Main Street
Anytown, USA 00000
(555) 555-1234

Working Conditions	Plus Percentage	Minus Percentage
Weather		
Crew Skills/Supervision		
Type of Work/ Degree of Difficulty		
Size of Job/Economy of Scale		
Site Congestion		
Inspections/Specs		
Working Hours/Overtime		
Distance to Stocking Pile		
Assembled/Unassembled Material		
Mounting Heights		
Work Space		
Total Adjustment Percentage		

Download this form at **www.DeWALT.com/guides**

Estimating Example

It normally takes 3 hours to set a small transformer. But now you are estimating a difficult job and decide 25% needs to be added to everything to cover possible productivity loss due to site conditions. Therefore, the adjusted man-hour for that transformer should be $3 \times (100\% + 25\%) = 3.75$ hours for one transformer installation.

JOBSITE MAN-HOUR WORKSHEET

ABC Contracting
1 Main Street
Anytown, USA 00000
(555) 555-1234

Job Name: _____   Ref No. _____

Superintendent: _____   Date: _____

Worksheet Number: _____

Item	Daily Output	Crew Member	Crew Hours	Total Man-Hours	Man-hour Per Unit

Download this form at **www.DeWALT.com/guides**

When recording man-hour information, specify the following:

- Materials and installation methods (e.g., dimensions, types)
- Type of jobs (e.g., new or renovation, residential or nonresidential)
- Crew (e.g., one worker or a team, including supervision and apprentice)
- Tools and equipment (e.g., scaffolding, lift)
- Weather conditions (e.g., rain, snow, wind, and temperature)

Check the jobsite at the same time every morning and record the number of units (such as the volume of concrete poured, the area of wall framed) that were installed the previous day. Repeat the same procedures on different projects over a period of time to verify result.

CONVERTING MINUTES TO DECIMAL HOURS

Minutes	Decimal in Hours	Minutes	Decimal in Hours
1	0.017	31	0.517
2	0.033	32	0.533
3	0.050	33	0.550
4	0.067	34	0.567
5	0.083	35	0.583
6	0.100	36	0.600
7	0.117	37	0.617
8	0.133	38	0.633
9	0.150	39	0.650
10	0.167	40	0.667
11	0.183	41	0.683
12	0.200	42	0.700
13	0.217	43	0.717
14	0.233	44	0.733
15	0.250	45	0.750
16	0.267	46	0.767
17	0.283	47	0.783
18	0.300	48	0.800
19	0.317	49	0.817
20	0.333	50	0.833
21	0.350	51	0.850
22	0.367	52	0.867
23	0.383	53	0.883
24	0.400	54	0.900
25	0.417	55	0.917
26	0.433	56	0.933
27	0.450	57	0.950
28	0.467	58	0.967
29	0.483	59	0.983
30	0.500	60	1.000

ESTIMATING LABOR BURDEN RATE

Estimating Math

Labor Burden Rate = Total Labor Burden/Total Basic Wages × 100%

Estimating Example

Annual Wages	Amount
John	$ 54,000
Adrian	$ 40,000
Tom	$ 35,000
Jack	$ 28,000
Wages subtotal	**$157,000**

Labor burdens	
Bonus	$ 8,000
Living Allowances	$ 1,000
Social Security	$ 9,734
Medicare	$ 2,277
Federal Unemployment Tax (FUTA)	$ 224
State Unemployment Tax (SUTA)	$ 720
Worker's Compensation Insurance	$ 9,813
General Liability Insurance	$ 1,523
Health Insurance	$ 24,000
Dental Insurance	$ 4,800
Pension (401K)	$ 9,420
Union Dues	$ 1,440
Labor burden subtotal	**$ 72,951**

Thus, Labor Burden Rate = Labor Burden/Wage = $72,951/$157,000 = 46.5%.

SALARY CONVERSION TABLE

Per Hour	Per Week	Per Month	Per Year
$ 6.00	$ 240	$1,039	$12,470
$ 7.00	$ 280	$1,212	$14,549
$ 8.00	$ 320	$1,386	$16,627
$ 9.00	$ 360	$1,559	$18,706
$10.00	$ 400	$1,732	$20,784
$11.00	$ 440	$1,905	$22,862
$12.00	$ 480	$2,078	$24,941
$13.00	$ 520	$2,252	$27,019
$14.00	$ 560	$2,425	$29,098
$15.00	$ 600	$2,598	$31,176
$16.00	$ 640	$2,771	$33,254
$17.00	$ 680	$2,944	$35,333
$18.00	$ 720	$3,118	$37,411
$19.00	$ 760	$3,291	$39,490
$20.00	$ 800	$3,464	$41,568
$21.00	$ 840	$3,637	$43,646
$22.00	$ 880	$3,810	$45,725
$23.00	$ 920	$3,984	$47,803
$24.00	$ 960	$4,157	$49,882
$25.00	$1,000	$4,330	$51,960
$26.00	$1,040	$4,503	$54,038
$27.00	$1,080	$4,676	$56,117
$28.00	$1,120	$4,850	$58,195
$29.00	$1,160	$5,023	$60,274
$30.00	$1,200	$5,196	$62,352
$31.00	$1,240	$5,369	$64,430
$32.00	$1,280	$5,542	$66,509
$33.00	$1,320	$5,716	$68,587
$34.00	$1,360	$5,889	$70,666
$35.00	$1,400	$6,062	$72,744
$36.00	$1,440	$6,235	$74,822
$37.00	$1,480	$6,408	$76,901
$38.00	$1,520	$6,582	$78,979
$39.00	$1,560	$6,755	$81,058
$40.00	$1,600	$ 6,928	$ 83,136

Per Hour	Per Week	Per Month	Per Year
$41.00	$1,640	$ 7,101	$ 85,214
$42.00	$1,680	$ 7,274	$ 87,293
$43.00	$1,720	$ 7,448	$ 89,371
$44.00	$1,760	$ 7,621	$ 91,450
$45.00	$1,800	$ 7,794	$ 93,528
$46.00	$1,840	$ 7,967	$ 95,606
$47.00	$1,880	$ 8,140	$ 97,685
$48.00	$1,920	$ 8,314	$ 99,763
$49.00	$1,960	$ 8,487	$101,842
$50.00	$2,000	$ 8,660	$103,920
$51.00	$2,040	$ 8,833	$105,998
$52.00	$2,080	$ 9,006	$108,077
$53.00	$2,120	$ 9,180	$110,155
$54.00	$2,160	$ 9,353	$112,234
$55.00	$2,200	$ 9,526	$114,312
$56.00	$2,240	$ 9,699	$116,390
$57.00	$2,280	$ 9,872	$118,469
$58.00	$2,320	$10,046	$120,547
$59.00	$2,360	$10,219	$122,626
$60.00	$2,400	$10,392	$124,704
$61.00	$2,440	$10,565	$126,782
$62.00	$2,480	$10,738	$128,861
$63.00	$2,520	$10,912	$130,939
$64.00	$2,560	$11,085	$133,018
$65.00	$2,600	$11,258	$135,096
$66.00	$2,640	$11,431	$137,174
$67.00	$2,680	$11,604	$139,253
$68.00	$2,720	$11,778	$141,331
$69.00	$2,760	$11,951	$143,410
$70.00	$2,800	$12,124	$145,488
$71.00	$2,840	$12,297	$147,566
$72.00	$2,880	$12,470	$149,645
$73.00	$2,920	$12,644	$151,723
$74.00	$2,960	$12,817	$153,802
$75.00	$3,000	$12,990	$155,880

NOTE

This table is based on a standard 40-hour week, 4.33-week month of a 52-week year. It does not consider vacation or overtime pays, or any other additional factors.

TIPS FOR ESTIMATING AND IMPROVING LABOR PRODUCTIVITY

Labor is quite difficult to estimate and even harder for job cost control. The following commonsense tips may be useful to you.

For estimators:

- Study plans and specs carefully.
- Know what your crew can handle.
- Get familiar with tool and equipment inventory.
- Divide unfamiliar tasks into smaller components to estimate.
- Keep good records of actual field performance.

For project managers or superintendents:

- Work out a realistic schedule and stick to it.
- Have submittals approved in a timely manner.
- Have right materials on site in the right quantities at the right time.
- Make tools and equipment available on the job when needed.
- Have drawings accurately marked up, including all documentation.
- Track job progress and crew productivity properly.
- Train better workers through company supported education.
- Have the home office provide timely administrative support to the field.
- Improve communications with subtrades.
- Maintain a clean, safe, and drug-free worksite.

ESTIMATING EQUIPMENT COSTS

If you are a trade contractor, there is no doubt you will have to include the jobsite overhead costs such as equipment for the work. If you are a general contractor self-performing some work, however, should you leave the equipment costs as part of total markup for the overall project? Not necessarily.

Suppose you rented a mobile crane onsite. If it is used primarily for your self-performed work, such as framing, then include crane cost (rental and operation) under "Framing." This makes sense, because that piece of equipment is part of the costs to do framing work. By including this crane under framing cost, your markup for the total project will then look smaller. Then the owner will have fewer questions and less chances of attacking your fees.

Similar examples include: vibrator for concrete work, foundation layout, forklifts, small tools, equipment hauling, concrete or framing foreman, travel expenses, and truck rental. As long as they are used for self-performed work, you can include them in the costs for the work itself, rather than as part of the general overhead. If the owner questions your tactics, you can explain, "Equipment is part of the cost to do the work."

Estimating Math

$$\text{Equipment Costs} = \text{Equipment Hours} \times (\text{Equipment Rate} + \text{Operator Rate})$$

Estimating Example

A pump is used to pour 18 cy of concrete footing. Normally, within an hour, that pump can pour 6 cy.

Equipment hours are 18/6 = 3 hours.
Pump rate is $100 per hour and operator rate is $40 per hour.
Equipment cost is 3 × (100 + 40) = $420

NOTES

1. Operator rate can be figured in a similar way as labor pricing:

   Operator Rate = Operator Basic Wage × (1 + Labor Burden Rate)

2. In figuring the equipment hourly rate, consider associated costs such as depreciation and interest (if financed), taxes and licensing, insurance, storage, tires, fuel, lubricants and filters, repair services. If the equipment is rented, normally you only have to pay for rental and fuel. But check your rental agreement to confirm what is covered and what is not.

PRICING SUMMARY WORKSHEET

The following worksheet has worked quite well for trade contractors. If you are a general contractor self-performing some work, use this as a guide. Remember to include some markup for the work itself.

ABC Contracting
1 Main Street
Anytown, USA 00000
(555) 555-1234

Item	Rate	Amount	Subtotal
Material Quotes			
Sales Tax/Freight			
Material Subtotal			
Labor Subtotal including burdens			
Subcontractor Costs			
Equipment Costs			
Other Related Jobsite Overhead			
Total Direct Costs			
Office Overhead for this Work			
Owner's Allowance for this Work			
Bond/Insurance for this Work			
Work Permit/License for this Work			
Contingency for this Work			
Profit for this Portion of Work			
Total Price for this Work			

CHAPTER 4

EVALUATING QUOTES

Evaluating price quotes from subtrades or suppliers is one of the most important estimating activities. Use your time effectively and do things right the first time. Simply plugging in quotes is not the correct way to estimate a project.

Try to get quotes in writing, by fax, mail, or email. When time is an issue, take the quote over the phone and keep written records.

In this chapter, the following topics will be covered.

- General principles for evaluating quotes
- Telephone quotation worksheet
- Solved quote evaluation examples
- Detailed guidelines and worksheets for every subtrade from earthwork to electrical

EVALUATING SUBCONTRACTOR QUOTES

Read quotes from subcontractors carefully before deciding on a low number. You should have at least three quotes for each trade. To compare quotes "apples to apples", first define a list of adjustment factors, that is, what you plan to add or deduct from the quotes. These factors might come from your own familiarity with the job, or from studying one of the "better-looking" quotes. Next, adjust each quote to reveal the low bidder.

The following questions may be helpful to you.

- Did you solicit quotes from this subcontractor?
- How much do you know about this subcontractor?
- Is the subcontractor also bidding to your competition?
- Is the quote in accordance with plans and specs?
- Is the subcontractor using specified material?
- Did the subcontractor review and acknowledge all addenda?
- Is the subcontractor aware of schedule requirements and liquidated damage clause?
- What are the exclusions and inclusions?
- Is the subcontractor bondable? What is the cost to provide a bond?
- Did the subcontractor break down the quote in the format as required?

• Did the subcontractor make any math errors in the quote?

SUBCONTRACTOR EVALUATION WORKSHEET

ABC Contracting
1 Main Street
Anytown, USA 00000
(555) 555-1234

Subtrade	Plug QTY/Price	Sub 1	Sub 2	Sub 3
Base Price				
Adjustments				
1				
2				
3				
4				
5				
Adjusted Price				
Notes				

Download this form at **www.DeWALT.com/guides**

Subcontractors can be evaluated in the same way as material suppliers, Yet material suppliers may prepare their quotes from a close takeoff, whereas subcontractors may have different inclusions and exclusions. Thus, it is important to pay attention to their scopes of work.

Apparently, with adjustment factors, the subcontractor who first appears to be the low bidder may actually be high. More importantly, after a careful evaluation process, you may catch items a sub missed or excluded, thus covering yourself from potential profit losses.

Example Subcontractor Evaluation

ABC Contracting
1 Main Street
Anytown, USA 00000
(555) 555-1234

Trade Quoted	Plug Quantity/Price	Sub 1	Sub 2	Sub 3
Communication System				
Base Price		$15,000	$14,000	$16,000
Work Included	Telephone	Yes	Yes	Yes
	Intercom	No	No	No
	Cable	Yes	Yes	Yes
	Internet	Add $2,000	Add $2,000	Incl.@ $2,000
Delivery Time		N/A	N/A	N/A
Adjustments				
Sales Tax		Included	Add $840	Included
Delivery to Jobsite		Included	Included	Included
Complete Installation		Included	Included	Included
Per Plans and Specs		Yes	Yes	Yes
Exclusions		None	None	None
Addenda Received		Yes	Yes	Yes
Other Factors		N/A	N/A	N/A
Adjusted Total Bid		**$17,000**	**$16,840**	**$16,000**

TELEPHONE QUOTATION WORKSHEET

ABC Contracting
1 Main Street
Anytown, USA 00000
(555) 555-1234

Date and time: _____

Project name: _____

Quote taken by: _____

Trade: _____ will follow up with a written quote: ___Yes ___No

Subcontractor/supplier name: _____

Contact person: _____

Telephone: _____ Fax: _____

Addenda received: _____

Scope of work: _____

Proposal amount: _____

Alternates: _____

Inclusions: _____

Exclusions: _____

Sales tax: ___Yes ___No

Delivery to jobsite: ___Yes ___No

Installation: ___Yes ___No

Per plans and specs: ___Yes ___No

Adjusted total amount: _____

Download this form at **www.DeWALT.com/guides**

Frequently telephone quotations come at the last minute of the bid. The following worksheet can be used to record important details of telephone quotes.

EXAMPLE 1: USING YOUR OWN YARDSTICK

When evaluating prices for an important trade, it is a good idea to measure up subs' quotes with your own "plug" price. Even if you do not self-perform, an accurate estimate can be done by measuring quantities and carefully applying unit prices using historical cost data.

This example shows a bid tab of sitework quotes. You use your own estimate to compare all quotes by different sections, including earthwork (including demo), paving (including curbs, signage, and line painting), utilities (e.g., storm, sanitary, water, fire, and reuse water, gas, electrical conduits), surveying and layout, mobilization, general conditions, and profit.

For adjustment factors, the basic rule is: *If the item is covered elsewhere by another trade, then take it out from this trade; otherwise you need to make sure everyone in this trade has it.*

In this example, site concrete sidewalk was shown on drawings but the building concrete sub's quote does not have it, so the site subs must include it. Grease traps happened to be included in the plumber's proposal, so you can remove it from the site subs' quote.

It seems Sub B is low, but instead of the $818,584 quoted, the actual price could be $775,796, which means a bidding advantage for you. Meanwhile, your own estimate gives you some confidence in this exercise.

Description		Your Estimate	Sub A	Sub B	Sub C	Sub D
Base Bid		$903,473	$888,619	$818,584	$930,818	$885,250
Breakdown						
	Earthwork	$132,314	$94,518	$96,131	$77,977	$53,400
	Paving/ Signage/Curb	$505,361	$466,098	$452,254	$438,117	$481,850
	Drainage	$138,440	$157,389	$138,074	$175,551	$167,202
	Sewer	$62,333	$83,480	$30,479	$101,658	$75,798
	Water	$24,539	$25,485	$16,362	$105,810	$85,000
	Fire line	$40,486	$61,650	$70,288	INCL	$22,000
	Survey/ As-Builts	INCL	INCL	$11,500	INCL	INCL
	Mobilization/G.C.	INCL	INCL	$3,500	$31,705	INCL
Adjustment						
Add	Demo	INCL (@$8, 640)	INCL	INCL	INCL	INCL
Add	Dewatering	$27,500	INCL (@ 30,000)	$27,500	INCL	INCL (@$25,000)
Add	Concrete Sidewalk	INCL (@$23,936)	INCL (@$24,670)	INCL (@$15, 900)	$21,502	$21,502
Deduct	Soil Testing	Not Included	($14,000)	No	No	No
Deduct	Fire Line	($40,486)	($61,650)	($70,288)	($61,650)	($85,000)
Deduct	Grease Trap	No	($8,800)	No	No	No
Adjusted Bid		$890,487	$804,169	$775,796	$890,670	$821,752

Preparing a Plug Estimate for Subcontracted Work

Use the following steps to estimate the subcontractor's work.

1. Carefully study drawings to define a complete scope of work. For example, to estimate drywall, are there any parapet metal framings even if the exterior wall is precast concrete?

2. Read through related spec sections to find out the quality of work. For example, how many coats of paint are needed, and what are inspection requirements?

3. Measure quantities off drawings (be generous and also be accurate).

4. Apply unit prices to each item. For example, for each interior wall assembly, apply individual unit prices for each square foot of gypsum board, batt insulation, and metal framing.

5. Summarize all items to reach a total plug-in price for trade evaluation purposes later.

The key for this type of estimating is to work out an accurate unit price for each item. You should watch subtrade prices close enough for a certain period of time to have a good grip of numbers, and then decide the best rate to use according to current job conditions. The following is an example for estimating plug numbers.

Estimating Example

<table>
<tr><td colspan="5" align="center">**ABC Contracting**
1 Main Street
Anytown, USA 00000
(555) 555-1234</td></tr>
<tr><th>Description</th><th>QTY</th><th>Unit</th><th>U/P</th><th>Total</th></tr>
<tr><td>**Earthwork**</td><td></td><td></td><td></td><td></td></tr>
<tr><td>Clear and grubbing</td><td>11</td><td>acre</td><td>$2,000</td><td>$22,000</td></tr>
<tr><td>Import fill</td><td>1775</td><td>cy</td><td>$6.50</td><td>$11,538</td></tr>
<tr><td>Proof roll</td><td>53250</td><td>sy</td><td>$0.50</td><td>$26,625</td></tr>
<tr><td>Rough grade</td><td>53250</td><td>sy</td><td>$0.65</td><td>$34,613</td></tr>
<tr><td>Building pad</td><td>8171</td><td>sy</td><td>$2.75</td><td>$22,471</td></tr>
<tr><td>Place and compact</td><td>1775</td><td>cy</td><td>$0.70</td><td>$1,243</td></tr>
<tr><td>Silt fence</td><td>2881</td><td>lf</td><td>$1.80</td><td>$5,186</td></tr>
<tr><td>Demo concrete sidewalk</td><td>4320</td><td>sf</td><td>$2.00</td><td>$8,640</td></tr>
<tr><td>**Earthwork Total**</td><td></td><td></td><td></td><td>**$132,314**</td></tr>
<tr><td></td><td></td><td></td><td></td><td></td></tr>
<tr><td>**Paving**</td><td></td><td></td><td></td><td></td></tr>
<tr><td>Subbase 12" stabilized</td><td>25308</td><td>sy</td><td>$2.20</td><td>$55,678</td></tr>
<tr><td>8" Limerock</td><td>25308</td><td>sy</td><td>$6.20</td><td>$156,910</td></tr>
<tr><td>Asphalt 1–1/2"</td><td>19324</td><td>sy</td><td>$5.50</td><td>$106,280</td></tr>
<tr><td>Asphalt 2"</td><td>5985</td><td>sy</td><td>$6.40</td><td>$38,302</td></tr>
</table>

(continues)

Description	QTY	Unit	U/P	Total
Prime and sand paving	25308	sy	$0.50	$12,654
Concrete apron 6"	5274	sf	$4.25	$22,415
12" Subbase for concrete apron	586	sy	$2.20	$1,289
Misc concrete sidewalk	553	sf	$2.75	$1,521
Signs and line painting				
Stop sign	11	ea	$210.00	$2,310
Other traffic regulation signs	3	ea	$125.00	$375
No parking sign	4	ea	$125.00	$500
Handicap sign	8	ea	$125.00	$1,000
Paint parking stall	364	ea	$5.00	$1,820
Paint handicap stall	8	ea	$35.00	$280
Paint stop bar	17	ea	$20.00	$340
Paint arrows	55	ea	$15.00	$825
Paint only message	4	ea	$40.00	$160
4" White solid traffic stripe	1200	lf	$0.20	$240
24" Thermoplastic bar	3	ea	$400.00	$1,200
Curbs	7401	lf	$13.50	$99,914
Wheelstops	19	ea	$25.00	$475
Handicap ramp	5	ea	$175.00	$875
Paving Total				**$505,361**
Utilities				
Storm Drainage				
C Inlets	12	ea	$1,250.0	$15,000
Tie-in manhole	2	ea	$750.0	$1,500
12" RCP	300	lf	$25.0	$7,500
15" RCP	300	lf	$25.0	$7,500
18" RCP	233	lf	$27.5	$6,408
24" RCP	285	lf	$37.5	$10,688
30" RCP	150	lf	$47.5	$7,125
42" RCP	248	lf	$70.0	$17,360
60" RCP	195	lf	$90.0	$17,550
MES	2	ea	$625.0	$1,250
Roof drain tie-in	13	ea	$600.0	$7,800

(continues)

Description	QTY	Unit	U/P	Total
Clean-out	3	ea	$300.0	$900
12" PVC	878	lf	$30.0	$26,340
8" PVC	480	lf	$24.0	$11,520
Storm Drainage Subtotal				**$138,440**
Sanitary Sewer				
Tie-in	2	ea	$700.0	$1,400
Manholes	8	ea	$1,875.0	$15,000
8" PVC	1716	lf	$20.0	$34,320
6" PVC	541	lf	$18.0	$9,738
4" PVC	125	lf	$15.0	$1,875
Sanitary Sewer Subtotal				**$62,333**
Water				
1 1/2" PVC	675	lf	$10.0	$6,750
2" PVC	215	lf	$11.0	$2,365
4" PVC	538	lf	$23.0	$12,374
Tie-in	5	ea	$490.0	$2,450
Testing and balance	1	ea	$600.0	$600
Water Subtotal				**$24,539**
Fire underground				
DDCV	1	ea	$6,400	$6,400
FDC	1	ea	$1,200	$1,200
FH Ass GV/Box	3	ea	$2,500	$7,500
6" Gate Valve	2	ea	$85.0	$170
4" PVC	285	lf	$8.0	$2,280
6" PVC	1305	lf	$10.2	$13,246
8" PVC	765	lf	$12.0	$9,180
Tees	6	ea	$85.0	$510
Fire underground subtotal				**$40,486**
Utilities Total (drainage, sanitary, water, fire)				**$265,798**
Site Work Total (earthwork, paving, utilities)				**$903,473**

The following shows how to determine unit price for 2 inch asphalt paving per square yard (not including base gravel or subbases). You established a job cost database for many trade items, and this table is taken from the database. The 2 inch asphalt paving ranges from $4.28 to $6.37 per square yard. You can take an average of all numbers or simply use the high end number (say, $6.40). Judging from the prices of 3 inch and 4 inch paving, the $6.40 seems safe.

Item	Price	Unit	Cost Source
Asphalt Paving 2"	$ 5.95	sy	ABC School
Asphalt Paving 2"	$ 6.37	sy	XYZ Hospital
Asphalt Paving 2"	$ 4.28	sy	Sigma Univ
Asphalt Paving 2"	$ 6.37	sy	Holiday Motel
Asphalt Paving 2"	$ 6.21	sy	Econ Golf Club
Asphalt Paving 2"	$ 6.29	sy	Cheap Grocery
Asphalt Paving 3"	$ 6.21	sy	Econ Golf Club
Asphalt Paving 3"	$ 8.22	sy	Green Mall
Asphalt Paving 3"	$ 8.72	sy	Cheap Grocery
Asphalt Paving 4"	$ 9.56	sy	ABC School
Asphalt Paving 4"	$ 14.58	sy	Holiday Motel
Asphalt Paving 4"	$ 11.69	sy	Econ Golf Club

It is important to update the cost database whenever possible. Use your judgment if a new unit price obtained seems to be unusually high or low.

EXAMPLE 2: COORDINATING DIFFERENT TRADES
Consider the Following Situation:

You received drywall quotes from three subs for a small commercial building:

- Sub A quoted interior drywall $21,500, exterior drywall $39,000. The exterior price included furnishing and installing preengineered metal trusses. Sub A also included interior/exterior wood blocking and plywood.
- Sub B quoted interior drywall $18,500, exterior drywall $29,000. Sub B included installing preengineered metal trusses furnished by others. Sub B excluded any wood blocking or plywood.
- Sub C quoted interior drywall $20,000, exterior drywall $25,000. Sub C has no trusses, no wood, or plywood. Sub C was only willing to add $1,800 to put plywood sheathing over trusses furnished by others.

All three prices include furnishing, hanging and finishing drywall and metal stud framing, batt insulation, scaffolding, equipment, and cleanup.

You also have quotes from metal truss manufacturers. The lowest material-only price is $12,500, including the shipping and taxes.

Your detailed rough carpentry estimate is as follows:

- Interior wood blocking and plywood sheathing for metal framing: $500
- Interior wainscot wood base: $700

- Exterior wood blocking for metal framing: $800
- Exterior truss plywood sheathing: $2,000
- Labor only to install trusses: $9,500

Suggested Solution:

To determine the low drywall price, start with a proposal that appears to be more complete (Sub A in this case) and find out what the quote includes. Then adjust other subs' quotes by adding items they did not include to match Sub A

Scope	Sub A	Sub B	Sub C
Interior			
Drywall	$21,500	$18,500	$20,000
Wood/plywood	INCL	Add $500	Add $500
Wainscot base	INCL	Add $700	Add $700
Interior Total	**$21,500**	**$19,700**	**$21,200**
Exterior			
Drywall	$39,000	$29,000	$25,000
Truss material	INCL	Add $12,500	Add $12,500
Truss labor	INCL	INCL	Add $9,500
Plywood on truss	INCL	Add $2,000	Add $1,800
Wood blocking	INCL	Add $800	Add $800
Exterior Total	**$39,000**	**$44,300**	**$49,600**
Total Drywall Price	**$60,500**	**$64,000**	**$70,800**

In this worksheet, the lowest sub appears to be Sub A for the total drywall prices. The combination of using Sub B for interior and Sub A for exterior seems to be lower, although subs may be unwilling to sign contracts for interior or exterior work alone. You can also ask Sub A to deduct an amount for trusses if you plan to buy these directly from the manufacturer. Finally, to be more competitive, review your rough carpentry estimate and deduct for wood blockings and plywood sheathing covered by the drywall sub.

EXAMPLE 3: EVALUATING "COMBO" QUOTES

Consider the Following Problem:

Seven subs are bidding part of the job with drywall, acoustical ceiling, and VCT.

- Sub 1 will do drywall and acoustical ceiling for $52,000.
- Sub 2 will do drywall only for $34,000.
- Sub 3 will do acoustical ceiling and VCT for $20,000.

- Sub 4 will do drywall, acoustical ceiling, and VCT for $60,000.
- Sub 5 will do acoustical ceiling for $15,000.
- Sub 6 will do VCT for $9,000.
- Sub 7 will do acoustical ceiling for $13,000.

Suggested Solution:

Quotes from subs 1, 3, and 4 are called "combo" quotes, because these subs perform a variety of trades for one combined price. You should ask for revised quotes with separate numbers for each trade, although subs may be uncooperative as they want to quote the work as a total package. In such cases, you may take the following steps.

1. Make a matrix, including all scopes of work for each sub.
2. For every sub, plug in the lowest price for anything he missed. In this example, the lowest stand-alone price for acoustical ceiling is $13,000, $9,000 for VCT, $34,000 for drywall. By doing so, you are making every sub "jack of all trades".
3. Add up all numbers for each sub, including the plug-in prices.

Sub	Drywall	Ceiling	VCT	Total
1	$52,000	Included	Add $9,000	**$ 61,000**
2	$34,000	Add $13,000	Add $9,000	**$ 56,000**
3	Add $34,000	$20,000	Included	**$ 54,000**
4	$60,000	Included	Included	**$ 60,000**
5	Add $34,000	$15,000	Add $9,000	**$ 58,000**
6	Add $34,000	Add $13,000	$9,000	**$ 56,000**
7	Add $34,000	$13,000	Add $9,000	**$ 56,000**

Now you will see clearly the lowest option is Sub 2 for drywall and Sub 3 for acoustical ceiling and VCT.

EXAMPLE 4: CORRECTING ERRORS

Subs frequently make mistakes in their quotes. Some are simple math errors, others are from incomplete scopes and careless quantity takeoff. The best way to deal with errors is to contact the sub and ask for a revised quote. Ironically, you may have to ensure that the sub is also sending corrected quotes to your competitors, if he is bidding to them as well.

Often, you will be unable to get a revision on time, especially when the errors were because of imperfections in plans and specs. For example, different quantities of plants are shown on the landscaping drawings from the ones indicated on the plant schedule. Some subs are quoting according to the plant schedule, while others are counting trees off the drawings.

If you are running out of time, decide what your pricing is based on (normally go with the conservative side), and adjust the sub's quotes accordingly. For example, you decided there should be 36 red maple trees and the sub quoted 34. Simply scan the quote for the price of red maple trees (say, $800 each). Add the difference $800 \times (36 - 34) = $1,600 to the base bid, then clearly note the discrepancy in the "list of bid clarifications" to owner.

EXAMPLE 5: DEALING WITH ALTERNATES

In many jobs the owner requests pricing for using alternative methods or substitute materials. The alternates are generally listed on the proposal form and may either add or deduct to your base bid. Because the owner could award the job by evaluating your base bid plus all alternates, do not ignore the pricing of alternates. You must obtain certain alternates from the subs, so ensure they understand the requirements.

The following guidelines may help when determining alternates.

- Think about the scope of the project. If you take something out, then you must replace it with something else. For example, if you decrease storefront height, then you must add more wall area using masonry, drywall, or cladding.

- Be careful about voluntary alternates. Your sub may submit an alternate that is not required. For example, say your lowest concrete sub's base price is $23,000 with an alternate add of $2,000 for using 4,000 psi instead of 3,000 psiI concrete for beams and columns. But the drawings require 4,000 psi concrete. The alternate price required for the bid is demolishing the existing building, which has nothing to do with the quote. Therefore, you should use $25,000, not $23,000, for concrete.

- Have your own estimate for the alternates, because subs may omit or be unwilling to price alternates.

- Make sure your final alternate prices include some markup. If the alternate is an add, then figure a percentage (normally 10% to 15%) to be added to the sub's number. If the alternate is a deduct, then do not take further deductions from the sub's number but present it as is.

The fear of adding too little or deducting too much is always a concern in pricing alternates. For example, Sub A's base bid is $10,000 with an alternate $5,000 add, and Sub B's base bid is $8,000 with an alternate $9,000 add. You decided to use Sub B, but at the last minute you receive a new quote from Sub C. The base bid is $7,000 with an alternate $10,000 add. Before deciding who and what to use, contact all subs to determine the basis of their prices. At the very least, ensure that alternates come from the same sub as used for the base bid. For example, if you decide to use Sub C, then use his alternate price as well.

Estimating Example

Alternate 1: Add 100 ft of 6 ft high screen wall

Trade	Price
Concrete	$ 8,000
Masonry	$ 3,000
Precast wall cap	$ 3,000
Stucco	$ 3,000
Paint	$ 1,200
Subtotal	**$ 18,200**
Add overhead & profit (10%)	$ 1,820
Alternate 1 Add	**$ 20,020**

Alternate 2: Use VCT in lieu of ceramic tile

Base Bid	QTY	UNIT	U/P	SUBTOTAL
Floor tile	350	sf	$ 12.00	$ 4,200
Wall tile	950	sf	$ 15.00	$ 14,250
Total				**$ 18,450**
Alternate				
Floor VCT	350	sf	$ 4.00	$ 1,400
Wainscot	950	sf	$ 5.00	$ 4,750
Total				**$ 6,150**
Alternate 2 Deduct				**$(12,300)**

EVALUATING SITEWORK QUOTES

Evaluation Questions

General

1. Is mobilization cost included? Does the site sub need multiple mobilizations?
2. Does the plan call for certified survey and/or as-built?
3. Is the site permit cost included if required?
4. Is the site material testing cost included if required?
5. Does a temporary access road need to be built?

Demolition

1. Is there demolition involved, exterior or interior?
2. Are there existing buildings to be totally demolished? Is there hazardous material to be removed? Is there an environmental report available for existing building?
3. Is there building interior selective demolition? How will that be coordinated with mechanical and electrical trades? Is temporary shoring and support required? Who is patching floors?
4. Is temporary fence or partition included to protect the demolition?
5. Who will remove the trash?

Earthwork

1. How many cubic yards of dirt are being moved? Is the site a cut or fill?
2. Will the excavated dirt be reused, piled onsite, or hauled away? Is trash or tree burning allowed?
3. Are there existing trees to be removed or relocated?
4. How long will it take for the site sub to get the building pad ready?

5.   Did the sub review the soil report? Is dewatering required?

6.   Are there hazardous materials present? Was this an existing landfill?

7.   Are there retention ponds, new or existing, shown on the plans?

8.   Is roadwork involved?

9.   Who is covering excavation for building foundation and slab prep?

Utilities

1.   Break down quotes into smaller portions such as storm drainage, sanitary sewer, water, and fire. Are there any gas line or telephone conduits? What work will be done by utility companies?

2.   Is a lift station included in sewer price? Is the electrician providing the power for lift station? Any septic tanks?

3.   Are the exterior connections to building plumbing, fire, and electrical systems included?

4.   Is site fire underground system included? Who is licensed to install fire line, site sub, or fire sprinkler sub?

5.   Are water meter fees and installation costs included? Is the water system chlorinization included, if applicable?

Paving

1.   Is line painting and signage included? Concrete curbs?

2.   Is paving asphalt or concrete? Who is providing base and subbase?

3.   Is concrete flatwork included, e.g., sidewalks or driveway aprons?

4.   Does the quote include traffic signalization as well as maintenance of traffic?

Miscellanies

1.   Are there jacking, boring, and piling to install irrigation sleeves, telephone, or electrical conduits?

2.   Is there a retaining wall or screen wall? Are those walls made of cast-in-place concrete, precast concrete, or masonry?

3.   Are there monument or pylon signs? Any details?

4.   Are there guardrails, chain link fence, or metal/wood fence?

5.   Are there site furnishings such as brick pavers, benches, bicycle racks, trash receptacles, tree grates?

6.   Are there site features such as gazebos or arbors? How are they constructed and finished?

Sitework Evaluation Worksheet

ABC Contracting
1 Main Street
Anytown, USA 00000
(555) 555-1234

Sitework Subs	Plug Price	Sub 1	Sub 2	Sub 3
Base Price				
Mobilization				
Site demolition				
Building demolition				
Site access				
Site permit				
Survey and as-builts				
Earthwork				
Soil preload				
Jacking, boring, and piling				
Rapid impact compaction				
Shoring and engineering				
Dirt removal				
Rock blasting				
Silt fence				
Dewatering				
Foundation excavation and backfill				
Slab gravel base				
M/E excavation and backfill				
Asphalt paving and gravel base				
Pit-run subbase				
Concrete curbs				
Curb prep and backfill				
Concrete sidewalk				
Site signage				
Line painting				

(continues)

Sitework Subs	Plug Price	Sub 1	Sub 2	Sub 3
Storm drainage				
Sanitary sewer				
Water system				
Fire underground				
Lift station				
Grease/oil trap				
Water meters				
Water chlorinization				
Utility connects to building				
Backflow preventers				
Soil testing				
Material testing				
Site retaining/screen wall				
Guardrails				
Chain link fence				
Monument signs				
Site furnishings				
Hazardous materials				
Offsite roadwork				
Open cut and repair				
Traffic maintenance				
Adjusted Price				

Download this form at **www.DeWALT.com/guides**

EVALUATING LANDSCAPING/IRRIGATION QUOTES
Evaluation Questions

1. Are landscape and irrigation included in base bid, or does the owner specify a cash allowance?
2. Do the quantities shown on the drawings match the plant schedule?
3. Are contractors quoting the right type of plant (e.g., different gallons of plants will vary in price)?
4. Is the top soil for plants included?
5. What are the maintenance requirements for the plants?
6. Are existing trees to be relocated, protected, or removed? Are there tree grates or guards?

7. Are sod, mulch, and seed included in the quote? Does the quote have enough quantities? (Some sods might not show on plans but are required, such as sods along the roadway.)

8. Are irrigation drawings available? Did the sub quote according to plans, or is the quote based on design-build?

9. Where is the water source? (Watch for irrigation wells or pumps.)

10. Is an irrigation backflow preventer included in the quote?

11. Should irrigation sleeves be included?

12. Is water meter installation for irrigation included? Are water meter fees included?

Landscaping/Irrigation Evaluation Worksheet

ABC Contracting
1 Main Street
Anytown, USA 00000
(555) 555-1234

Landscaping Subs	Plug Price	Sub 1	Sub 2	Sub 3
Base Price				
Trees and shrubs				
Sod, seed, and mulch				
Top soil				
Remove existing trees				
Maintenance				
Specified material				
Irrigation system				
Wells or pumps				
Backflow preventer				
Irrigation sleeves				
Irrigation power				
Water meters				
Planter drain				
Tree grates				
Tree protection				
Brick pavers with base				
Site furnishings				
Adjusted Price				

Download this form at **www.DEWALT.com/guides**

EVALUATING FORMWORK/CONCRETE QUOTES
Evaluation Questions

1. Does the building require auger piling? If yes, you need a separate quote.

2. Do the plans call for any special concrete work such as tilt-up panels, precast architectural concrete, lightweight concrete? If yes, you need separate quotes for those. Make sure engineering and shop drawing costs are included.

3. Is the formwork sub furnishing concrete? Or is it labor only?

4. Is the formwork sub furnishing and installing rebar?

5. Does the quote include concrete beam and columns (vertical work)? Or is it foundations and slabs (flat work) only?

6. Does the quote include exterior building sidewalk and dumpster pad?

7. Does the quote include the concrete needed for plumbing, HVAC, electrical such as equipment pad, electrical conduits, and plumbing trenches?

8. Who is constructing the site concrete sidewalk, driveway, and foundation for the retaining wall?

9. Does the quote include fine grading for building slab?

10. Does the quote include soil termite treatment, if applicable?

11. Does the quote include miscellaneous material for concrete work, such as wood blocking and insulation for freezer slab, joint sealers for slab on grade, chamfer strips for columns and walls?

Formwork/Concrete Evaluation Worksheet

ABC Contracting
1 Main Street
Anytown, USA 00000
(555) 555-1234

Formwork/Concrete	Plug Price	Sub 1	Sub 2	Sub 3
Base Price				
Forming				
Concrete material				
Concrete placing				
Concrete finishing				
Rebar supply or install				
Excavation/backfill				
Fine grading				
Soil treatment				
Footings/pads				
Foundation walls				
Slab on grade				
Suspended slabs				
Upper floor topping				
Beams and columns				
Shaft walls				
Stairs				
Building sidewalks				
Site concrete				
Layout				
Scaffolding				
Forklift/crane				
Expansion joints				
PVC water stop				
Chamfer strip				
Form openings				
Patching exposed wall				
M/E equipment pads				
Embeds				
Rebar unloading				
Adjusted total				

EVALUATING REINFORCING QUOTES
Reinforcing Evaluation Worksheet

ABC Contracting
1 Main Street
Anytown, USA 00000
(555) 555-1234

Rebar Subs	Plug Price	Sub 1	Sub 2	Sub 3
Base Price				
Rebar for foundation				
Rebar for slab				
Wire mesh for slab				
Masonry dowels				
Exterior concrete reinforcing				
Grade of steel				
Galvanized				
Epoxy coating				
Sales tax				
Delivery				
Rebar unloading				
Rebar cutting				
Rebar install				
Quantity (weight)				
Adjusted Price				

Download this form at **www.DeWALT.com/guides**

EVALUATING MASONRY QUOTES
Evaluation Questions

1. How many blocks or bricks did the mason figure? Obtain this information to compare with your own numbers.
2. Does the quote include cell fill concrete (material and labor)?
3. Does the quote include masonry rebar (material and labor)?
4. Does the quote include scaffolding and equipment?
5. Are there any fire resistant requirements for the blocks?
6. Does the quote include bricks, if any? What is the unit price for brick supply, if not owner's cash allowance?
7. If the quote says pre-ast is included, does it mean architectural precast concrete or structural precast lintels?
8. Is cast stone or stone veneer, if shown, included in the quote?
9. Are there dumpster enclosures or site block walls?
10. Does the quote include rigid/foam insulation or vapor barrier?
11. Does the quote include miscellaneous items, such as flashing, precast sills, waterproofing, installation of embedded metals (e.g., brick support angle), and pressed steel door frames?

Masonry Evaluation Worksheet

ABC Contracting
1 Main Street
Anytown, USA 00000
(555) 555-1234

Masonry Subs	Plug Price	Sub 1	Sub 2	Sub 3
Base Price				
CMU				
Brick				
Stone				
Glass blocks				
Site masonry				
Rebar supply				
Rebar install				
Cell fill concrete				
Mortar and sand				
Scaffolding				
Forklift				
Install door frame				
Precast lintel installed				
Arch precast installed				
Wall cleaning				
Masonry bracing				
Brick flashing				
Caulking				
Air/vapor barrier				
Foam insulation				
Fire resistance				
Adjusted Price				

Download this form at **www.DeWALT.com/guides**

EVALUATING STRUCTURAL STEEL/MISCELLANEOUS METALS QUOTES

Evaluation Questions

1. How many tons of steel are included in the quote?

2. Does the quote include both fabrication and erection? If they are coming from different companies, do their scopes match?

3. In the material portion of the quote, are taxes and freight costs included?

4. Ensure that the following miscellaneous steels are included for both material and labor:

   - Handrails and guardrails
   - Stairs
   - Ladders
   - Steel lintels
   - Dumpster gates and posts
   - Steel bollards
   - Framing for HVAC units
   - Supporting angles for interior partitions
   - Supporting steel for exterior building signage
   - Rainhoods over exterior doors

5. Are shop drawings signed and sealed by professional engineers? If yes, what is the additional cost for this?

Steel/Miscellaneous Metals Evaluation Worksheet

ABC Contracting
1 Main Street
Anytown, USA 00000
(555) 555-1234

Metal Subs	Plug Price	Sub 1	Sub 2	Sub 3
Base Price				
Erection				
Sales tax				
Delivery				
Engineered shop drawing				
Prime/paint				
Joist and decking				
Beams and columns				
Structural wall bracing				
Metal canopy				
Metal stairs and railing				
Interior corridor handrail				
Balcony guardrail and privacy screen				
Elevator hoist and divider beams				
Elevator access ladder				
Roof ladder				
RTU angle support framing				
Roof parapet steel angle				
Exterior wall grilles				
Exterior retaining wall guardrail				
Exterior metal gates				
Foundation embedded angle				
Column base plates and anchor bolts				
Window steel lintels				
Galvanized bollards				
Masonry veneer support angle				
Overhead door framing				
Rainhoods over doors				
Column protectors				
Trench drain frame and grating				
Garbage/laundry chute channel				
Adjusted Price				

Download this form at **www.DEWALT.com/guides**

EVALUATING FRAMING/CARPENTRY QUOTES
Evaluation Questions

1. Is this project using stick framing or preengineered floor joists and trusses?

2. Are there fire rated or pressure treated requirements for lumber?

3. Is heavy timber or wood decking present?

4. Are drywall contractors including rough carpentry items for their scope of work? Adjust prices accordingly.

5. Are millwork quotes from approved manufacturers listed?

6. Is wood blocking for installing millwork included?

7. Are countertops solid surface or plastic laminate?

8. Does the quote include material and labor for any fiber reinforced panels (FRPs), vinyl ceilings, cornice trims, crown moldings, wood bases?

9. Does the quote include labor for installing interior doors and hardware, toilet accessories?

10. Are there furniture, fixtures, and equipment furnished by the owner and installed by the contractor? If yes, figure support blockings and labor.

Framing Evaluation Worksheet

Framers	Plug Price	Sub 1	Sub 2	Sub 3
ABC Contracting 1 Main Street Anytown, USA 00000 (555) 555-1234				
Base Price				
Framing material				
TJI/truss supply				
Lumber/plywood supply				
Laminated beams				
Floor framing and sheathing				
Wall framing and sheathing				
Roof framing and sheathing				
Fascia and soffit				
Framing layout				
Nails/Screws				
Power				
Forklift				
Crane for joist/truss				
Backframing				
Wood stairs				
Adjusted Price				

Download this form at **www.DEWALT.com/guides**

Millwork Evaluation Worksheet

ABC Contracting
1 Main Street
Anytown, USA 00000
(555) 555-1234

Millwork Subs	Plug Price	Sub 1	Sub 2	Sub 3
Base Price				
Approved manufacturer				
Installation				
Warranty				
Kitchen cabinets				
Bathroom countertops				
Medicine cabinets				
Bath tub surrounds				
Plastic laminate toilet partitions				
Shelving or display cases				
Wood wall paneling				
Laboratory casework				
Custom furniture				
Doors and gates				
Millwork hardware				
Adjusted Price				

Download this form at **www.DEWALT.com/guides**

Finish Carpentry Evaluation Worksheet

ABC Contracting
1 Main Street
Anytown, USA 00000
(555) 555-1234

Finish Carpenters	Plug Price	Sub 1	Sub 2	Sub 3
Base Price				
Exterior wood trims				
Exterior vinyl soffit				
Interior wood baseboards				
Cornice trims				
Crown molding				
Window sills				
Door casings				
Wall edge casing				
Corridor wood handrail				
FRP panels				
Wood wall paneling				
Supply of trim materials				
Installation of doors				
Installation of door frames				
Installation of door hardware				
Installation of washroom accessories				
Installation of owner's FFE				
Adjusted Price				

Download this form at **www.DeWALT.com/guides**

EVALUATING ROOFING QUOTES

Evaluation Questions

1. How many types of roof system are present (flat, metal, tile, shingle)?

2. Are roofers quoting the approved system? Check specs.

3. Are they approved installers of the specified system?

4. Is roof insulation included? What is the R-value?

5. Are gutters and down sprouts made of stainless or aluminum?

6. How are roof penetrations done? Are roof hatches included in the quote? Are skylights plastic or aluminum?

7. If existing roof, does the quote include repairing of the old roof?

8. How many years of warranty are roofers quoting, labor warranty and material warranty? What is required?

9. Are roofers including aluminum rainlock panels, if any? (These may be found on the rear side of the roof parapet.)

10. Does the quote include testing costs for the roofing system?

11. Does the quote include walking pads or concrete roof pavers?

12. Does the quote include metal sidings on exterior wall?

Roofing Evaluation Worksheet

ABC Contracting
1 Main Street
Anytown, USA 00000
(555) 555-1234

Roofers	Plug Price	Sub 1	Sub 2	Sub 3
Base Price				
Supply per specs				
Approved installer				
Sales tax				
Flat roof				
Sloped roof				
Metal canopy				
Sidings and claddings				
Roof insulation				
Vapor barrier				
Downsprouts/gutters				
Flashing				
Roof drain/scupper				
Roof pavers				
Rainlock panels				
Skylights				
Roof hatch with post				
Repairing existing roof				
Membrane on exposed parkade				
Plywood roof sheathing				
Roof system testing				
Special warranty				
Adjusted Price				

Download this form at www.**DeWALT**.com/guides

EVALUATING INSULATION/WATERPROOFING QUOTES

Evaluation Questions

1. Are drywall and acoustical ceiling subcontractors furnishing batt insulation for their scope of work?

2. Is there any rigid insulation along the perimeter of the building and under the slab? What about behind brick veneer or EIFS?

3. Did the mason include foam insulation for blocks?

4. Is the concrete floor to be sealed? Are the expansion joints to be caulked? Did painters include caulking?

5. Is there an elevator pit? Do you have quotes on waterproofing? (The exterior foundation wall may also need to be water sealed or damp proofed.)

6. Is fireproofing required? Check both architectural and structural drawings as well as specs for the extent of spray-on fireproofing.

7. Mechanical and electrical contractors often include fire stopping for their scope of work (if mentioned in their spec sections). You may need to allow additional money for fire stopping that is not M/E related.

Insulation/Waterproofing Evaluation Worksheet

ABC Contracting
1 Main Street
Anytown, USA 00000
(555) 555-1234

Insulation Sub	Plug Price	Sub 1	Sub 2	Sub 3
Base Price				
Batts to sloped roof and attics				
Batts to flat roof				
Batts to roof patios				
Batts to exterior wall				
Batts to party and corridor wall				
Batts to shaft and sound wall				
Batts to other interior partitions				
Batts above sound ceiling				
Insulation to wood framed floor edge				
Batts fill to floor overhang				
Poly vapor barrier to walls				
Poly vapor barrier under slabs				
Spray on cellufibre insulation				
Rigid Insulation				
Firestopping				
Sprayed fireproofing				
Rod and caulking to windows				
Adjusted Price				

Download this form at **www.DeWALT.com/guides**

EVALUATING DECKS/RAILINGS QUOTES

Residential projects often require aluminum glazed guardrails for balcony decks or roof patios. The subcontractor who specializes in this work also will cover the vinyl waterproof membrane for the deck.

Decks/Railings Evaluation Worksheet

ABC Contracting
1 Main Street
Anytown, USA 00000
(555) 555-1234

Deck Subs	Plug Price	Sub 1	Sub 2	Sub 3
Base Price				
Balcony deck				
Balcony flashing				
Balcony railing				
Glazing to railing				
Balcony privacy screen				
Post wraps				
Building interior railing				
Adjusted Price				

Download this form at **www.DEWALT.com/guides**

EVALUATING DOOR QUOTES

Evaluation Questions

1. How many doors are included in the quote? Do a door count yourself.
2. Are doors premachined and prefinished? Or do they require painting?
3. How many types of doors are shown? Note special doors, such as overhead door, will be quoted from separate contractors, not hollow metal door suppliers.
4. Are those doors from specified manufacturers?
5. Did door suppliers include material taxes in their proposals? What about freight?
6. Is hardware package included in the price? Is it per specs?
7. Are there any fire resistant requirements for the doors?
8. Who is installing those doors? If there are a lot of doors, did finish carpenters include the installation of doors in their quotes?

Hollow Metal and Wood Doors Evaluation Worksheet

ABC Contracting
1 Main Street
Anytown, USA 00000
(555) 555-1234

Door Suppliers	Plug Price	Sub 1	Sub 2	Sub 3
Base Price				
Supply per specs				
Sales tax				
Hollow metal doors				
Wood doors				
Prehung wood doors				
Suite entry doors				
Bifold doors				
Mirrored bipass doors				
Pocket doors				
Patio doors				
Screen doors				
Pressed steel frames				
Pocket door frames				
Wood door frames				
Pressed steel window frames				
Finish hardware				
Card readers				
Mag locks				
Glass lites on doors				
Installation				
Adjusted Price				

Download this form at **www.DEWALT.com/guides**

Special Doors Evaluation Worksheet

ABC Contracting
1 Main Street
Anytown, USA 00000
(555) 555-1234

Special Doors	Plug Price	Sub 1	Sub 2	Sub 3
Base Price				
Supply per specs				
Sales tax				
Installation				
Automatic entrance doors				
Revolving doors				
Overhead doors				
Grilles				
Vertical lift doors				
Traffic doors				
Accordion folding doors				
Hardware				
Glazing for doors				
Auto door opener				
Adjusted Price				

Download this form at **www.DeWALT.com/guides**

EVALUATING STOREFRONT/WINDOWS QUOTES
Evaluation Questions

1. Is there a list of approved manufacturers for curtain wall system?
2. Is there a list of approved manufacturers for glass and glazing?
3. Is glass impact resistant? Are hurricane shutters or storm panels required?
4. Are automatic doors quoted separately? Who is providing glass for automatic doors?
5. Does the quote include glass lites on steel doors?
6. Is there any glass inside the building?
7. What about windows? (Obtain a separate quote for residential vinyl windows, as they are normally supply only.)
8. Are there any large mirrors? Are they included under glass or specialties? Are they in custom sizes?
9. Should glass contractors provide signed and sealed shop drawings?
10. Is glass cleaning included in the quote?
11. Does the quote include caulking for glazing and aluminum mullions?

Glass and Glazing Evaluation Worksheet

ABC Contracting
1 Main Street
Anytown, USA 00000
(555) 555-1234

Glass Subs	Plug Price	Sub 1	Sub 2	Sub 3
Base Price				
System per specs				
Sales tax				
Installation				
Curtain wall				
Spandrel panels				
Storefront				
Aluminum windows				
Glass doors				
Automatic sliding doors				
Shower doors				
Patio sliding doors				
Washroom mirrors				
Glazing to interior windows				
Glazing to interior partitions				
Glazing to hollow metal doors				
Hardware for aluminum doors				
Handicap operators				
Electric strikes				
Impact resistance				
Site glazing				
Glass cleaning				
Caulking				
Engineered shop drawings				
Adjusted Price				

Download this form at **www.DeWALT.com/guides**

Vinyl Window Supply Evaluation Worksheet

ABC Contracting
1 Main Street
Anytown, USA 00000
(555) 555-1234

Window Suppliers	Plug Price	Sub 1	Sub 2	Sub 3
Base Price				
Installation				
Sales tax				
Freight				
Engineered shop drawing				
Exterior windows				
Patio sliding doors				
Interior windows				
Flashing				
Liners				
Reinforcing				
Screen				
Aluminum guards				
Site glazing				
Adjusted Price				

Download this form at **www.DEWALT.com/guides**

EVALUATING DRYWALL, STUCCO, AND ACOUSTICAL CEILING QUOTES

Evaluation Questions

Drywall

1. Does the quote include both drywall and framing?
2. Does the quote include batt insulation for drywall? Does it include all types of drywall required, such as moisture resistant drywall (green board) and furring on concrete/masonry walls?
3. For fire protection requirements, does the interior demising wall stop right at the ceiling or go to the underneath of the roof deck?

4. Does the quote include any rough carpentry items (e.g., wood blocking/plywood sheathing)?

5. Does the quote include scaffolding and equipment?

6. If this is a renovation project, does it involve patching or repairing existing drywall surfaces?

7. Does the quote include installation of hollow metal door frames?

8. Are signed and sealed shop drawings required for metal framing?

9. If the quote includes preengineered trusses, are they wood or metal? (Make sure both materials and labor costs are included.)

10. Does it include other trades (stucco, acoustical ceiling)?

EIFS and Stucco

1. Does this quote include both EIFS and stucco?

2. Does the quote include metal lath to support stucco?

3. Did the subcontractor include any drywall?

4. Does the quote include special finish items such as Fypon trim, foam shapes, or glass reinforced fiber columns (GRFCs)?

5. Does the quote include scaffolding and equipment?

Ceiling

1. What type of ceiling is required (e.g., drywall, T-bar, metal, wood, exposed metal joist and decking)? Sometimes there is no reflected ceiling plan, but the finish schedule might specify the ceiling.

2. What type are acoustical panels (e.g., 2×2 or 2×4)?

3. Is suspension system included? What about egg crates?

4. Is batt insulation for ceiling required?

5. Are there acoustical panels on the wall?

6. Are cut-outs for sprinkler heads included?

7. Are there special ceiling-hung light fixtures (e.g., chandeliers) to be installed? (Normally, these are supplied by the owner and the electrician will only provide the wiring.)

Drywall/Stucco/Ceiling Evaluation Worksheets

ABC Contracting
1 Main Street
Anytown, USA 00000
(555) 555-1234

Drywall/Stucco/Ceiling Subs	Plug Price	Sub 1	Sub 2	Sub 3
Base Price				
Drywall				
Textured ceiling				
Ceiling bulkheads				
Acoustical ceiling with suspension				
Acoustical wall panels				
Stucco with metal lath				
EIFS and foam shapes				
Exterior trims				
GFRC columns				
Metal framing				
Metal truss supply and install				
Furring on concrete/masonry wall				
Densglass sheathing				
Batt Insulation				
Rigid insulation				
Vapor barrier				
Corner bead				
Aluminum reveals				
Wood blocking				
Plywood sheathing				
Scaffolding				
Hoisting				
Install interior door frames				
Install access doors				
Cut and patching				
Engineered shop drawing				
Adjusted Price				

Download this form at **www.DeWALT.com/guides**

EVALUATING FLOORING QUOTES
Evaluation Questions

1. How many different types of flooring are in this bid (e.g., hard tile, VCT, sheet vinyl, carpet, hardwood, laminate, quartz, resinous)? Within each category of flooring, further identify each type (e.g., ceramic tile differs from porcelain tile or marble tile, even though they all could be called "hard tile").

2. Does the quote include floor preparation?

3. Is the base included with flooring (e.g., the rubber base for VCT and carpet, the vented base for wood flooring)?

4. Is flooring installed on the wall (e.g., ceramic tile on restroom walls, in showers, and on building exterior face)?

5. Do you have separate quotes for special flooring, such as access flooring?

6. If a renovation project, is the removal of existing floor included? What about floor patching?

7. Is floor cleaning cost included (e.g., the sealing and waxing of VCT, vacuuming of carpet)?

Flooring Evaluation Worksheet

ABC Contracting
1 Main Street
Anytown, USA 00000
(555) 555-1234

Flooring Subs	Plug Price	Sub 1	Sub 2	Sub 3
Base Price				
Carpet				
Carpet on wall				
Sheet vinyl				
VCT				
Wood flooring				
Floor tile				
Ceramic tile on wall				
Flooring bases				
Special flooring				
Floor preparation				
Remove existing flooring				
Threshold				
Transition strip				
Clean, seal, wax				
Adjusted Price				

Download this form at **www.DeWALT.com/guides**

EVALUATING PAINTING QUOTES

Evaluation Questions

1. Did the quote cover both interior and exterior painting?

2. Are there vinyl wall coverings that shall be included? Is the owner furnishing the material? Are they custom made?

3. Are doors to be stained or painted? Are there wood trims to paint, exterior and interior?

4. Are the millwork cabinets prefinished or to be painted?

5. For renovation projects, is the new painting to match existing building? Is existing building to be repainted?

6. Are there any special acrylic waterproofing coatings on exterior masonry blocks or precast concrete wall panels?

7. Does the quote include any line painting for parking?

8. Is the underneath of exposed structural steel metal deck to be painted?

9. What is the finishing requirement for miscellaneous metals such as railing and bollard?

10. Does the quote include scaffolding and equipment?

Painting Evaluation Worksheet

ABC Contracting
1 Main Street
Anytown, USA 00000
(555) 555-1234

Painting Subs	Plug Price	Sub 1	Sub 2	Sub 3
Base Price				
Per plans and specs				
Exterior painting				
Interior painting				
Vinyl wall coverings				
Paint columns and beams				
Paint doors				
Paint steel handrails				
Paint wood trims				
Paint or prime to ceilings				
Paint stairs				
Paint parking lot				
Paint underground parkade				
Paint existing surface				
Special wall coating				
Painting Subs	**Plug Price**	**Sub 1**	**Sub 2**	**Sub 3**
Caulking				
Scaffolding				
Adjusted Price				

Download this form at **www.DEWALT.com/guides**

EVALUATING SPECIALTIES QUOTES

Evaluation Questions

1. Is the owner to furnish and/or install these items?

2. Who are the approved manufacturers?

3. Are installation costs included in the quote? What about freight and taxes?

4. Does the installation only involve loose labor, or require mechanical and electrical connection? What about the costs for site receiving, unloading, and storage? Is there any hoisting equipment to be used?

5. Are expensive items included, such as hand dryers, lockers and benches, flag poles, bike racks?

6. Is chain link fencing shown on site or building plans?

7. What type of pavers are quoted (e.g., concrete or brick, 4 × 4 or 4 × 8)? How many square feet of pavers? Any specific manufacturer? Are the pavers to be cleaned and sealed? What is the paver base (e.g., sand)?

8. For awnings, are they canvas or metal? If metal, are they steel or aluminum? Do they require signed and sealed drawings?

9. For swimming pools, what is the timeline for completion? Is the pool deck included? Is the sub including the excavation, concrete, finish, water supply, and electrical for the pool?

Specialties Evaluation Worksheet

ABC Contracting
1 Main Street
Anytown, USA 00000
(555) 555-1234

Specialties Suppliers	Plug Price	Sub 1	Sub 2	Sub 3
Base Price				
Approved manufacturers				
Sales tax				
Freight				
Installation				
Toilet partitions				
Urinal screens				
Washroom accessories				
Handryers				
Lockers				
Benches				
Corner guards				
Shower curtain and rods				
Shower doors				
Window blinds				
Closet wire shelving				
Residential appliances				
Fireplaces				
Bicycle racks				
Trash receptacles				
Mailboxes				
Flagpoles				

(continues)

Specialties Suppliers	Plug Price	Sub 1	Sub 2	Sub 3
Foot grilles				
Louvers and vents				
Chalkboards and tack boards				
Awnings				
Interior signage				
FFE				
Adjusted Price				

Download this form at **www.DeWALT.com/guides**

EVALUATING CONVEYING SYSTEM QUOTES
Evaluation Questions

1. How many floors in the building? What is the height of each floor? Are elevators required, and how many?
2. What type of elevator is designed, hydraulic or electrical?
3. Are they freight or passenger elevators?
4. Are there other conveying systems such as handicap lifts and chutes?
5. Who is the approved elevator manufacturer?
6. What is the loading capacity for elevator?
7. Are elevator inspection costs included in the quote?
8. Is the flooring in elevator cab included?
9. What is the lead time for elevator delivery after approval of shop drawings?
10. Is there coordination with other trades (e.g., hoist/divider beams with structural steel, elevator pit with concrete and reinforcing, power for elevator with electrical)?

Conveying Systems Evaluation Worksheet

ABC Contracting
1 Main Street
Anytown, USA 00000
(555) 555-1234

Elevator Subs	Plug Price	Sub 1	Sub 2	Sub 3
Base Price				
Approved manufacturers				
Permit and license				
Sales tax				
Freight				
Installation				
Hydraulic or electrical				
Handicap lift				
Dumbwaiter				
Garbage/linen chute				
Pneumatic tubing system				
Floor height				
Load capacity				
Elevator inspection				
Engineered shop drawings				
Lead time				
Flooring in cab				
Adjusted Price				

Download this form at **www.DEWALT.com/guides**

EVALUATING PLUMBING QUOTES

Evaluation Questions

1. Is plumbing permit included?

2. In coordinating with sitework, does the plumbing quote cover all the work to 5 ft outside the building? Is the connection with site water line covered?

3. Is roof drainage included? What about weeping tile?

4. Is excavation/backfill for underground work included?

5. Does the quote include all plumbing fixtures as required? Are fixtures quoted from approved manufacturers? Are large fixtures included, such as grease traps, water heaters, sump pumps?

6. If a renovation project, does it involve demolishing or modifying existing plumbing systems? Is concrete cutting and patching included?

7. Is there coordination with other mechanical work? Are such items as condensate drains covered, either by the HVAC sub or by the plumber?

8. Are costs for as-built drawings included?

9. Are water meters included?

10. Does the quote include concrete equipment pads?

Plumbing Evaluation Worksheet

ABC Contracting
1 Main Street
Anytown, USA 00000
(555) 555-1234

Plumbing Subs	Plug Price	Sub 1	Sub 2	Sub 3
Base Price				
Including HVAC/Fire sprinklers				
Plumbing permit				
Approved fixtures				
Water heaters				
Sump pump				
Grease trap				
Sand/oil interceptor				
Backflow preventers				
Roof drain				
Floor drain				
Condensate drain				
Trench drain				
Rainwater leader				
Weeping tile				
Pipe insulation				
Shower doors				
Coordinate with sitework				
Coordinate with HVAC				
Natural gas system				
Special piping				
Demolish existing system				
Cutting, coring, patching				
Washer/dishwasher connection				
Firestopping				
Excavation and backfill				
Water meters				
Temporary water				
Concrete equipment pads				
Seismic restraint				
As-built drawings				
Adjusted Price				

Download this form at **www.DeWALT.com/guides**

EVALUATING FIRE SPRINKLER QUOTES
Evaluation Questions

1. Is fire sprinkler system required? Are there any fire sprinkler drawings? (Also refer to reflected ceiling sheets. Check local building codes if no information is shown on drawings.)

2. Is the quote based on plans and specs, or simply design-build?

3. Is fire sprinkler permit included?

4. Is fire pump required? Is it included?

5. Coordinate with sitework. Does the sprinkler quote include any site fire underground lines? Does the quote cover all the work to 5 ft outside the building? Is the connection with exterior fire line covered?

6. For buildings that have exterior canopies or roof/floor overhangs, is a fire sprinkler system required at these locations?

7. Does the quote include costs for shop drawings and as-built drawings?

8. If this is a renovation project, does it involve demolishing or modifying existing fire sprinkler systems?

Fire Sprinklers Evaluation Worksheet

ABC Contracting
1 Main Street
Anytown, USA 00000
(555) 555-1234

Fire Sprinkler Subs	Plug Price	Sub 1	Sub 2	Sub 3
Base Price				
Per plans and specs				
Permit				
Wet or dry system				
Exterior canopy coverage				
Attic coverage				
Backflow preventers				
Fire pump				
Hangers and support				
Fire department connection				
Siamese connection				
Fire hose cabinets and racks				
Fire extinguisher and cabinet				
Coordinate with sitework				
Firestopping				
Demolish existing system				
Cutting, coring, patching				
Excavation and backfill				
Seismic restraint				
As-built drawings				
Adjusted Price				

Download this form at **www.DEWALT.com/guides**

EVALUATING HVAC QUOTES
Evaluation Questions

1. Is HVAC permit included? Is refrigeration system included?

2. Are rooftop HVAC units to be furnished by owner? Who will provide curbs and wood blocking for rooftop HVAC units? Is hoisting for the units included (e.g., cranes)?

3. Is HVAC equipment quoted from approved manufacturers?

4. Who is doing condensate drainage piping from the roof, the plumber or the HVAC contractor?

5. If this is a renovation project, does it involve demolishing or modifying existing HVAC system?

6. What kind of ductwork is included, sheet metal or fiberglass?

7. Does the quote include thermostats? Is the electrician including wiring for HVAC controls?

8. Are costs for shop drawings included?

9. Does the quote include certified test and balance?

10. Does the quote include concrete equipment pads?

HVAC Evaluation Worksheet

ABC Contracting
1 Main Street
Anytown, USA 00000
(555) 555-1234

HVAC Subs	Plug Price	Sub 1	Sub 2	Sub 3
Base Price				
Refrigeration				
Plumbing/fire sprinkler				
HVAC permit				
Approved equipment manufacturer				
Rooftop units				
Central A/C units				
P-TAC window units				
Curbs and wood blocking				
Equipment hoisting				
Condensate drain piping				
Heat pumps				
Electric heaters				
Sheet metal or fiberglass ductwork				
Ductwork insulation				
Gas/hot water piping, etc.				
Thermostats and control wiring				
Louvers and access panels				
Firestopping				
Coordinate with plumbing				
Coordinate with electrical				
Demolish existing system				
Cutting, coring, patching				
Certified test and balance				
Concrete equipment pads				
Seismic restraint				
Shop drawings				
As-built drawings				
Adjusted Price				

EVALUATING ELECTRICAL QUOTES

Evaluation Questions

Site Electrical

1. Does the quote include temporary power, including the power to hook up general contractor's trailer? Where is transformer location?

2. Who will provide the primary connection, that is, from power company main line to transformer?

3. Who will provide the secondary connection, that is, from the transformer into the building?

4. Is there site lighting involved in this bid? Are site light poles buried or mounted on the bases? Who will provide the bases as well as the poles and fixtures?

5. Is there any site underground telephone or electrical conduits? Are they to be concrete encased or covered with sand? Is excavation and backfill for these conduits included?

6. Who is providing power for site monument signs, if any?

7. Who is providing power for irrigation system?

8. Are utility company charges included in the quote?

Building Electrical

1. Is electrical permit included?

2. Are special systems, such as fire alarm systems, required, and included in the quote?

3. What about visual, audio, and data systems? (Sometimes the owner will provide the system but not the conduits.)

4. Are lighting fixtures and electrical panels to be supplied by the owner? Are there special requirements for lighting fixtures? Check both specs and interior design drawings.

5. Does the quote include the power for mechanical equipment and FFE hookup, especially for items supplied by the owner?

6. Is the excavation and backfill for underground conduits included?

7. Are there roof lights? Is the power for exterior building signage included?

8. If a renovation project, does it involve demolishing or modifying existing electrical systems? Is there concrete cutting and patching for that?

9. Does the quote include concrete equipment pads?

Electrical Evaluation Worksheet

ABC Contracting
1 Main Street
Anytown, USA 00000
(555) 555-1234

Electrical Subs	Plug Price	Sub 1	Sub 2	Sub 3
Base Price				
Electrical permit				
Temporary power				
Primary connection				
Secondary connection				
Approved lighting fixtures				
Electrical panels				
Utility company connection				
Site elect/phone/cable conduits				
Site lighting				
Light pole bases				
Excavation and backfill				
Fire alarm system				
Audio, visual and data systems				
Power for irrigation system				
Power for garage door				
Power for automatic door				
Power for mechanical equipment				
Power for FFE hookup				
Power for exterior/interior signage				
Power for elevator				
Lights on the roof				
Mechanical control wiring				
Access panels				
Firestopping				
Lamps for fixtures				
Coordinate with mechanical				

(continues)

Electrical Subs	Plug Price	Sub 1	Sub 2	Sub 3
Demolish existing system				
Cutting, coring and patching				
Concrete equipment pads				
Seismic restraint				
Shop drawings				
As-built drawings				
Adjusted Price				

Download this form at **www.DeWALT.com/guides**

CHAPTER 5

COST SUMMARY

Bid days can be quite hectic for those who prepare the price proposals and for those who receive them. No matter how ready you think you are, there are almost always surprises, if not problems. Being organized can help reduce errors and make the process less stressful.

In this chapter, the following topics will be covered.

- Estimating jobsite and home office overhead
- Estimating bond and insurance
- Estimating other indirect costs
- Estimating profit
- Cost summary worksheets
- Bid progress checklist
- Bid day workflow
- Writing a proposal and common proposal exclusions
- Assessing bid risks
- Cost breakdown requirements

ESTIMATING JOBSITE OVERHEAD

Jobsite overhead, or general conditions, is the money directly related to your job, including the costs to run a jobsite office and the salaries for jobsite personnel. Because jobsite overhead can amount to 20% to 40% of the total bid, it is necessary to estimate it with great accuracy. Do not apply a percentage to cover overhead. To make a thorough calculation, use the following guidelines.

- Define a list of jobsite overhead items you need to include.
- Decide how long it will take to finish the job. You may have to prepare a preliminary project schedule. Discuss the duration with your project managers or superintendents to find out what they think.
- Price each overhead item based on job duration. Note that the costs for some items are recurring on a regular basis throughout the job (e.g., daily, weekly, monthly, or yearly).
- Summarize all the costs to get a total jobsite overhead.

The following are common jobsite overhead cost items.

- Job mobilization and demobilization
- Salaries for project managers, superintendents, and foremen
- Travel expenses for field personnel
- Jobsite signs
- Trailer hookup, including temporary power
- Trailer monthly rental
- Field office furniture and supplies
- Monthly utilities (e.g., heat, electricity, gas, water, telephone)
- Small tools and tool sheds
- Material storage
- Temporary partitions, enclosures, fencing, gates, barricades
- Portable toilet rental
- Drinking water
- Safety equipment and first aid
- Drawing reproduction
- Postage for project documents
- Jobsite vehicles and fuels
- Parking for construction crew
- Scaffoldings
- Hoisting equipment
- Permits and licenses
- Surveying and layout
- Material testing costs for both site and building
- Miscellaneous cutting and patching
- General jobsite daily cleanup
- Final cleanup at completion including glass cleaning
- Monthly progress photos
- Allowance to correct punch-list items
- As-built drawings
- Operation and maintenance

The following shows the calculation for a four-month project.

ABC Contracting
1 Main Street
Anytown, USA 00000
(555) 555-1234

Item	QTY	Unit	Rate	Subtotal
Mobilization	1	l/s	$ 8,000	$ 8,000
Project manager	18	wk	$ 1,500	$ 27,000
Superintendent	18	wk	$ 800	$ 14,400
Project sign	1	l/s	$ 3,000	$ 3,000
Trailer rental	4	mon	$ 3,000	$ 12,000
Setup trailer	1	l/s	$ 2,500	$ 2,500
Field office furniture	1	l/s	$ 1,000	$ 1,000
Field office supplies	4	mon	$ 200	$ 800
Telephone bills	4	mon	$ 150	$ 600
Fax machine	1	l/s	$ 100	$ 100
Computers and IT support	1	l/s	$ 2,500	$ 2,500
Temporary power	4	mon	$ 4,000	$ 16,000
Water meters	1	l/s	$ 2,000	$ 2,000
Temporary water	4	mon	$ 800	$ 3,200
Portable toilets rental	18	wk	$ 450	$ 8,100
Safety equipment and first aid	1	l/s	$ 1,500	$ 1,500
Drawing reproduction	1	l/s	$ 2,000	$ 2,000
Postage	1	l/s	$ 500	$ 500
Vehicles and fuels	2	ea	$ 1,200	$ 2,400
Forklift	18	wk	$ 700	$ 12,600
Small tools	1	l/s	$ 900	$ 900
Perimeter temporary fence	500	w	$ 3	$ 1,500
Surveying and layout	1	l/s	$ 700	$ 700
Daily cleanup	90	day	$ 40	$ 3,600
Final cleanup	1	l/s	$ 400	$ 400
Punch-list items	1	l/s	$ 4,000	$ 4,000
As-built drawings	1	l/s	$ 1,300	$ 1,300
Total Jobsite Overhead				**$ 132,600**

ESTIMATING HOME OFFICE OVERHEAD

Home office overhead could not be directly tied to a specific job. You must pay for these costs to remain in business. Even if you are running your business from a pickup truck, some of the following could apply to you.

- Owner's salary
- Office personnel salaries and benefits (e.g., estimators, bookkeepers, secretaries)
- Vehicles not related to specific projects, with fuels and insurance
- Office rent, utilities, furniture, supplies
- Business license and membership dues
- Marketing and advertising
- Loan interest
- Legal and auditing expenses
- Taxes and donations
- Bad accounts

Office overhead varies annually. A good way to determine this is to review the amount of work you did last year and the overhead, figure a percentage based on that, and then apply that percentage to the current estimate. Remember, office overhead is a cost, not profit. It is better to calculate this amount separately from your profit, although some builders combine the two and call it "markup."

Estimating Math

Overhead Rate = Office Overhead Last Year/Construction Volume Last Year

Office Overhead for Current Job = Rate × Total Direct Costs for Current Job

Estimating Example

Home office overhead last year: $300,000

Construction volume last year: $5,000,000

Home office overhead rate: $300,000/$5,000,000 = 6%

Total direct costs for your current job: $1,900,000

Home office overhead for current job: $1,900,000 × 6% = $114,000

ESTIMATING BONDING COSTS

Some bids require bid bond to ensure you will sign the contract if the job is awarded to you. Your surety should provide bid bond for free, or for a small amount of annual service charge. Check in advance to see if a bid bond is required, so that you can get the paperwork done on time if necessary.

What might be added to the bid price is performance and payment bonds. A performance bond is your guarantee to the owner that you will finish the job according to contract documents. A payment bond, or labor and material bond, promises all labor and material supplied on the job will be paid for, protecting the owner from any claims. They are normally made out to 100% of the contract amount.

The expense for payment and performance bond is usually 1% to 3% of the total job cost. It can be calculated by using a rate table from your surety company.

The following shows the calculation based on a rate table for furnishing payment and performance bonds.

Construction Costs	Bond Rate/$1,000
First $100,000	$28.50
Next $400,000	$17.10
Next $2,000,000	$11.40
Next $2,500,000	$8.55
Next $2,500,000	$7.98
Over $7,500,000	$7.41

Suppose your job totals approximately $2.7 million. The bond cost is as follows:

First: $100,000: $100,000/$1,000 × $28.50 = $2,850

Next: $400,000: $400,000/$1,000 × $17.10 = $6,840

Next: $2,000,000: $2,000,000/$1,000 × $11.40 = $22,800

For the balance of construction costs:

($2,700,000 − $2,000,000 − $400,000 − $100,000)/$1,000 × $8.55 = $1,710

The total costs for payment and performance bonds is:

$2,850 + $6,840 + $22,800 + $1,710 = $34,200

A safer way to figure out the bonding costs, however, is to contact your surety company directly to get a price quote. By doing so, you can also ensure your surety is willing to bond the job.

Some large projects spanning several years may require maintenance bonds. In such cases, you will add more costs into the equation, based on job duration.

ESTIMATING INSURANCE COSTS

For estimating and bidding purposes, there are three major kinds of insurances you need to examine individually.

- Builder's risk insurance
- General liability insurance (can be part of the office overhead)
- Owner's wrap-up insurance (only in some cases)

Other types of insurances required could be included elsewhere. For example, it may be easier to include worker's compensation insurance in your home office overhead, while the automobile insurance for your jobsite vehicles could be put under jobsite overhead.

You must get a quote directly from your insurance company for builder's risk insurance. It protects the job against direct loss and the actual rate depends on job location and duration, type of construction, and your company history. When you have the quote, make sure the deductible is satisfactorily low, and it provides coverage from all possible hazards such as fire, lightning, wind, and flood.

Estimating Example

The construction value of a project is $20 million with duration of 2 years.

The builder's risk rate is $0.25 per $1,000 for each month.

The cost of builder's risk insurance is $0.25 × $20,000,000/$1,000 × 24 = $120,000.

ESTIMATING OTHER INDIRECT COSTS

Many builders are also developers and must pay for soft costs in addition to hard construction costs. Soft costs normally include the following:

- Land acquisition
- Architect and engineer design fees
- Construction interests
- Bank charges
- Legal fees
- Appraisal fees
- Accounting fees
- Building permit costs
- Development cost charges
- Land survey
- Property taxes
- New home warranty
- Strata fees
- Marketing and sales

To estimate soft costs, talk with people who are charging the fees. This is more reliable than making inaccurate guesses about the costs.

Estimating Example

Your city has the following fee schedule for building permit costs.

Construction Costs	Building Permit Costs
$1.00 − $1,000	$45
$1,001 − $100,000	$45 plus $8.50 per $1,000 or portion thereof
$100,001 − $500,000	$886.50 plus $7.00 per $1,000 or portion thereof
$500,001 and up	$3,686.50 plus $6.25 per $1,000 or portion thereof

Your city also charges the following development cost charges (DCC) per square foot of residential buildings.

Type	DCC per sf
Road	$1.06
Water	$0.07
Sewer	$0.30
Parks	$0.08

You are building a large custom home with the construction value of $2 million and area of 3,500 sf. The $2 million construction cost falls in the category, "500,001 and up."

Building permit cost is $3,686.5 + ($2,000,000 − $500,000)/$1,000 × $6.25 = $13,061.

Development cost charge is 3,500 × ($1.06 + $0.07 + $0.30 + $0.08) = $5,285. You can call the city building department to verify your calculations.

ESTIMATING PROFIT

Profit is the money you want to make from the job and is normally estimated by applying a rate to total costs. The rate could run 20% to 30% for small jobs and 10% to 15% for large jobs.

In deciding the rate to be used, look at the trend on your completed jobs. Keep a chart of completed jobs handy (as follows) and evaluate each job's actual performance against the estimate. Did you achieve the profit you desired very often? Then decide what sort of profit you can hope for on the current project.

ABC Contracting
1 Main Street
Anytown, USA 00000
(555) 555-1234

Job Name	Contract Amount	Name of Job Superintendent	As-Bid Profit Rate	Actual Profit Rate	Rate Evaluation
Today's Rate					

SUMMARIZING COSTS

Summarizing all costs into a correct total bid price is a crucial "make-break" moment, as any cost items you missed reduce overall profit. Following is a rundown of major cost areas.

- Self-performed work, including material, labor, equipment, markup
- Work by subcontractors
- Jobsite overhead (supervision, etc.)
- Office overhead
- Owner's cash allowance (check specs)
- Bond and insurance
- Permit and development fees
- Other soft costs (make a list)
- Contingency (money to cover bad design documents, material/labor escalation, unforeseeable field conditions)
- Profit

Keep in mind the best way to avoid errors is to prepare an estimate as detailed as possible.

Final Check

Some of the following items may be excluded from the subcontractor's quote. Allow additional money for anything that were not quoted.

- Foundation excavation and backfill
- Concrete finishing
- Miscellaneous metals
- Wood trims and window sills
- Roof flashing, gutters, downspouts
- Building perimeter weep tile
- Poly vapor barrier for exterior wall and under slab-on-grade
- Rigid insulation on wall and under slab
- Damp proofing and waterproofing
- Firestopping and fireproofing
- Caulking and sealants
- Labor for doors, frames, hardware and windows
- Labor for toilet accessories, appliances, cabinets, countertops
- Closet shelving
- Window blinds
- Access panels
- Corner guards
- Washroom mirrors and shower doors

BID ESTIMATE WORKSHEETS

With the deadline approaching, set up two estimate worksheets for the bid. First is the bid workup sheet, which is a list of trade components with their detailed costs. Second is a bid recap sheet, which summarizes all direct and indirect costs to get a bid total.

Bid Workup Sheet

To set up a bid workup sheet, review drawings and specs thoroughly to make a list of trade components as complete as possible. Do not simply list the spec sections, but be more detailed. Separate the components that require different pricing methods or quotations. For example, you may want to separate the "glass storefront" from "windows," because they could mean different trades.

A bid workup sheet should have the following basic elements.

- Item ID
- Item description
- Quantities
- Units
- Material unit price
- Material subtotal
- Labor unit price
- Labor subtotal
- Subcontractor unit price
- Subcontractor subtotal
- Total item costs
- Name of subcontractor or supplier

Make reasonable expectations about how you will obtain the price for each line item. Some items are self-performed, and for these you will plug in your own numbers (if possible, separate costs to show labor, material, and so on). For other items, you will get quotes from your subcontractors and suppliers, but a detailed takeoff is always helpful. For example, you can count doors and measure the area for stone veneer. Later you can compare your takeoff against subs' quotes to identify discrepancies, or use a plug number if you fail to get any quotes at all.

Bid Recap Sheet

The bid recap sheet shows the total bid price. Here you summarize the total for major cost categories, including material, labor, and subcontractor, and then add markups to reach a grand total bid price.

The focus of any bid day is to get a correct total price. If you set up your bid estimate in spreadsheet format, the bid total will be automatically generated and updated (works great for last-minute changes such as electrical).

BLANK BID WORKUP SHEET

ABC Contracting
1 Main Street
Anytown, USA 00000
(555) 555-1234

No.	Item	QTY	Unit	Labor U/P	Labor Subtotal	Material U/P	Material Subtotal	Subcon-tractor U/P	Subcon-tractor Subtotal	Item Subtotal	Sub/ Supplier's Name

Bid workup sheet math includes:

- Labor Subtotal = Quantities × Labor Unit Price
- Material Subtotal = Quantities × Material Unit Price
- Subcontractor Subtotal = Quantities × Subcontractor Unit Price
- Item Subtotal = Material Subtotal + Labor Subtotal + Subcontractor Subtotal

BLANK BID RECAP SHEET

The following is a sample recap sheet for commercial/institutional projects. For residential jobs, you may have more indirect costs to include.

ABC Contracting
1 Main Street
Anytown, USA 00000
(555) 555-1234

Description	Rate	Amount
Labor Total		
Material Total		
Subcontractor Total		
Total Item Costs		
Add Material Sales Tax	%	
Add Labor Burden	%	
Total Direct Costs		
Indirect Costs		
Overhead		
Bond		
Insurance		
Financing		
Permit		
Development Fees		
Total Indirect Costs		
Total Costs		
Profit	%	
Bid Total		

Bid recap sheet math includes:

- Labor Total = The sum of "Labor Subtotal" column on bid workup sheet
- Material Total = The sum of "Material Subtotal" column on bid workup sheet
- Subcontractor Total = The sum of "Subcontractor Subtotal" column on bid workup sheet
- Material Sales Tax = Material Total × Sales Tax Rate
- Labor Burden = Labor Total × Labor Burden Rate
- Total Direct Costs = Labor Total + Material Total + Subcontractor Total + Material Sales Tax + Labor Burden
- Total Indirect Cost = Overhead + Bond + Insurance + Financing + Permit + Development Fees
- Profit = (Total Direct Cost + Total Indirect Cost) × Profit Rate
- Bid Total = Total Direct Cost + Total Indirect Cost + Profit

BID PROGRESS CHECKLIST

One or two days before the bid, check how the bid is going, to avoid unpleasant surprises on the bid day. The following questions may be helpful in evaluating the situation.

- Does your bid workup sheet include everything required in the scope?
- If you copied the bid workup sheet from an old bid, have you made applicable changes according to the scope of this bid?
- Have you checked the math of all worksheets?
- Did architects respond to your questions?
- Have all addenda been issued and received? Have all affected subs and suppliers been notified?
- Have you completed detailed estimates for self-performed trades?
- Do you have enough coverage from qualified subs for important trades?
- Have you received the quotes for bonds from your surety?
- Have you received the quotes for builder's risk insurance?
- Have you found a room in the office where you will be working on the bid?
- Do you have a copy of complete drawings and specs with latest revisions?
- Have you prepared a bid proposal form to be submitted? Is it in spreadsheet format? Could it be linked to your bid estimate sheets?
- Can you get some of the documents to be submitted ready in advance (e.g., signatures, copy of business license, insurance certificates)?
- Does everyone in the office know that you are bidding the job on that day? Who will help you with the bid? Is there an extra fax machine available?
- Have you gathered and organized all the quotes in a folder?
- Have you checked the time, date, and place for the bid to be submitted?

After you work out a final dollar value but before turning in the bid, ask the following questions.

- Does my bid include everything required in the scope?
- Is my bid in accordance with plans and specs?
- Does my bid exclude everything that should not be included?
- Is my planned schedule realistic? Can I finish the job on time?
- Did I check the math from beginning to end?
- Did I break out the quote in the format as required by the bid?
- Are there gaps among subcontractors that I need to fill?
- Did I include the required owner's cash allowance?
- Did I allow contingencies for unclear designs, price increases, unforeseeable site conditions?
- Did I complete all blanks on the bid form, including alternate price?
- Is the bid proposal signed by the company owner?
- Have I checked the names of subcontractors specified on the bid form?
- Did someone else review the estimate and proposal?

BID DAY WORKFLOW

Assembling a successful bid usually involves team effort. Various trades are divided among team members for evaluating responsibilities.

1. Before starting to work on the bid, have a short meeting with your bid team. Review general scope of work by going through drawings, specs, and addenda. Discuss requirements such as cost breakdown, alternates, separate prices.

2. One hour before the bid is due, begin to take trade prices from your team members and put them into the bid workup sheet. You may discover problems in getting prices for certain trades. Ask your colleague to keep working on them until they are resolved.

3. Half an hour before the bid is due, be sure all prices are entered in the bid workup sheet. Resolve any problems with your team members immediately.

4. Ten minutes before the bid is due, check the bid workup sheet and the bid recap sheet to ensure the numbers match. Complete the bid proposal with the grand total number and fill all other blanks.

5. Five minutes before the bid is due, get approval from your supervisor or manager and submit the bid proposal form with all other required documents. Allow more time if the proposal is to be delivered in person.

COMPLETED BID WORKUP SHEET

ABC Contracting
1 Main Street
Anytown, USA 00000
(555) 555-1234

No	Item	QTY	Unit	Labor U/P	Labor Subtotal	Material U/P	Material Subtotal	Subcontractor U/P	Subcontractor Subtotal	Item Subtotal	Sub/Suppliers' Name
1	Sitework	1	l/s					$385,000	$385,000	$385,000	ABC Site
2	Landscaping	1	l/s					$180,000	$180,000	$180,000	XYZ Land Lease
3	Termite Control	25,000	sf			$0.08	$2,000			$2,000	ACME Termite
4	Building Concrete	1	l/s					$130,500	$130,500	$130,500	Adam Concrete
5	Building Sidewalk	1,000	sf					$4.00	$4,000	$4,000	Adam Concrete
6	Masonry	1	l/s					$175,000	$175,000	$175,000	Block Works
7	Stone Veneer	1,050	sf					$25.00	$26,250	$26,250	Block Works
8	Foam Insulation	1	l/s					$7,500	$7,500	$7,500	Northern Foam
9	Structual Steel	1	l/s					$145,000	$145,000	$145,000	Northwest Steel
10	Aluminum Trellis	1	l/s					$30,000.00	$30,000	$30,000	Dilworth Metals
11	Rough Carpentry	1	l/s	$25,000	$25,000	$15,000	$15,000			$40,000	General Contractor
12	Window sills	1	l/s					$4,500.00	$4,500	$4,500	Creative Millwork

(continues)

No.	Description	Qty	Unit								Vendor
13	Built-Up Roofing	1	l/s					$135,000	$135,000	$135,000	Countrywide de Roof
14	Caulking	1	l/s					$5,600	$5,600	$5,600	Caulking-All
15	H.M. Doors	3	ea	$125.00	$375	$650.00	$1,950			$2,325	Tom's Doors
16	Wood Doors	2	ea	$90.00	$180	$500.00	$1,000			$1,180	Tom's Doors
17	Hardware	5	ea	$50.00	$250	$300.00	$1,500			$1,750	Tom's Doors
18	Glass and Glazing	1	l/s					$52,000	$52,000	$52,000	Jerry's Glass
19	Windows	1	l/s					$7,560	$7,560	$7,560	Perfect Windows
20	Stucco	1	l/s					$65,000	$65,000	$65,000	Stucco Man
21	Drywall	1	l/s					$115,000	$115,000	$115,000	First-class Drywall
22	Acoustical Ceilings	1	l/s					$14,530	$14,530	$14,530	South Acoustics
23	VCT	1	l/s					$1,250	$1,250	$1,250	All Flooring
24	Painting	1	l/s					$38,000	$38,000	$38,000	Complete Painter
25	Toilet Accessories	1	l/s	$2,000	$2,000	$6,000	$6,000			$8,000	Pacific Supply
26	Awnings	1	l/s					$10,000	$10,000	$10,000	Sunroof Awning
27	Fire Protection	1	l/s					$28,000	$28,000	$28,000	Northern Fire
28	Plumbing	1	l/s					$9,000	$9,000	$9,000	Pandas Piping
29	HVAC	1	l/s					$74,580	$74,580	$74,580	Bernard Air
30	Electrical	1	l/s					$99,000	$99,000	$99,000	Gordon Electrical
31	Site lighting	1	l/s					$13,570	$13,570	$13,570	Richter Electrical

Please make sure each line item has a price (total item cost) and a name (the name of subcontractor/supplier). If you did not get a quote, use a plug number and put down your company name as the subcontractor. If you are using a supplier to provide the material and a sub for the labor, both names should be referred.

COMPLETED BID RECAP SHEET

ABC Contracting
1 Main Street
Anytown, USA 00000
(555) 555-1234

Description	Rate	Amount
Labor Total		$ 27,805
Material Total		$ 27,450
Subcontractor Total		$ 1,755,840
Total Item Costs		$ 1,811,095
Add Material Sales Tax	6%	$ 1,647
Add Labor Burden	25%	$ 6,951
Total Direct Costs		**$ 1,819,693**
Indirect Costs		
Home Office Overhead	10%	$ 181,969
Job Overhead		$ 135,000
Payment/Performance Bond		$ 40,000
Builder's Risk Insurance		$ 15,000
Financing Interest		$ 75,000
Permit		$ 15,000
Impact Fee		$ 20,000
Total Indirect Costs		**$ 481,969**
Total Costs		**$ 2,301,663**
Profit	8%	$ 184,133
Bid Grand Total		**$ 2,485,796**

If you set up the bid estimate in spreadsheet format, the above cost recap sheet will be automatically updated whenever there is a change in price.

WRITING A PROPOSAL

After figuring out a total bid price, you are now ready to submit a proposal. Owners normally have their proposal form you are required to use and follow.

If such forms do not exist and you are allowed to write your own price proposal, then choose the words carefully and include the following information.

- Proposal number, date, and version of revision
- Complete job name and address
- Names of owner, architect, engineers
- A complete list of drawings and their issue or revision date
- Specification titles and issue date
- A complete list of addenda and their dates of issue
- Base bid price and alternate prices
- A list of inclusions, exclusions, clarifications, assumptions
- Your contact name and phone number
- Valid Duration of this price proposal

Check the bid instructions for submission requirements. The owner may not allow you to fax or email bid proposal. But you may be able to send out the original copy by mail several days before the deadline and then fax a revision on the bid day.

BID PROPOSAL FORM

TO: School District #123

RE: All American School

FROM: ABC Contracting

Having carefully examined the drawings and specifications, visited the site, familiarized with bidding requirements and other factors affecting the costs, the undersigned submits proposals to furnish all labor and materials for constructing the above project as follows:

Base Bid (express in words and figures): Two Million Four Hundred Eighty-Five Thousand Seven Hundred Ninety-Six Only ($2,485,796) Dollars

Subject to the attached bid clarifications

Alternate 1: Use ceramic tiles in lieu of VCT, add $5,000

Alternate 2: Delete fire sprinkler system: deduct $20,000

Receipt of Addendum No(s) 1 and 2 is acknowledged.

This proposal is valid for 60 days from the date of the submission.

Signature:

Title: President

Date: January 1, 2009

COMMON PROPOSAL SCOPE EXCLUSIONS

Theoretically, a proposal should be "per plans and specs," but sometimes contract drawings are not very clearly designed. If allowed, you may have to list clarifications, detailing inclusions and exclusions in the price proposal.
   The following is a list of common proposal exclusions.

- Building permit
- Development cost charges
- Architectural or engineering design
- Any other type of soft costs
- Errors in design documents
- Independent inspections and testing
- Legal survey
- Major material price increases or wage escalations
- Costs for performance or payment bonds

- Delays or defaults due to strikes, accidents, fires, acts of God, or any other unavoidable causes beyond the contractor's reasonable control
- Reasonable access to site, including snow clearing
- Confined space work
- Handling of hazardous material (e.g., asbestos removal)
- Environmental testing
- Dewatering
- Overexcavation of soft spots and backfill of the same
- Importation of fill should native fill prove inadequate
- Utility connection charges
- Work related to underground obstructions or utilities beyond the scope of bid documents
- Building exterior signs and logo
- Food service equipment
- Furniture, fixtures, equipment except installing owner furnished items
- Nondestructive examination (x-ray, magnetic particle, ultrasound, etc.)
- Chemical cleaning
- Commissioning
- Extended warranty
- Landlord or tenant improvement work

Depending on the job scope, the items you need to exclude may differ from the previous list. For example, if you are doing the building only and the owner is hiring a contractor to do the sitework, then you are to determine whether the foundation earthwork is part of your estimate or not. If the answer is not clear, then make a detailed list of sitework items to be excluded (e.g., footing excavation and backfill, building pad preparation, slab subbase and base fill, landscaping, irrigation).

ESTIMATING RISK DOLLARS

You have completed your takeoff, entered the quotes, applied markup, and finished a clear proposal with inclusions and exclusions. You believe you have accounted for everything, so what could possibly go wrong? Lots! The question remains: Where will the project go wrong, and what can you do as an estimator to cover risks as much as possible?

Review the following factors.

- Duration of the project
- Number of labor and total hours required
- Potential hazards
- Weather conditions
- Potential delays caused by failure of other trades
- Possible procurement failures (e.g., owner-furnished fixture package)
- Quality of drawings and extent of change order expected
- Amount of engineering and drafting required
- Quantity of construction equipment and tools required

- Materials procurement cost
- Labor costs and its percentage in relation to the total price

The fundamental question is: Should you take this job? Refresh your memory as to why you decided to bid this job.

- Will the job disrupt or interfere with other operations of the company?
- Is the company trained and equipped for this kind of job?
- Is there a sufficiently trained labor force readily available?
- Is the present stock of tools and equipment adequate for this type of job?
- Are there other contractors available who are more able to handle this type of work?

Discuss the situation with experienced project managers and field superintendents. Review the job with them and listen to their thoughts. One way to compensate for the risks is to figure some risk dollars, which is how much it will cost you if things go wrong.

Estimating Example

Foundation is one area where recently you have lost money. You are currently bidding a small job (about $250,000 in total price). Foundation is approximately $30,000. Your typical loss is about 40%.

Your risk dollars for foundation is 40% × $30,000 = $12,000. Your revised total base bid is $250,000 + $12,000 = $262,000. The percentage of risk dollars in total bid is $12,000/$262,000 × 100% = 4.58%.

COST BREAKDOWN REQUIREMENTS

Due to financial needs, owners frequently require both a cost breakdown and a lump sump price. For example, if you are bidding two buildings in one bid, the owner may want to know the separate costs for each building. This breakdown, sometimes called a "schedule of values," could help the owner to understand the billings once the job starts.

A common basis is breaking out price by the gross building area (i.e., square feet). Suppose there are several buildings in one job. Normally the larger ones have a larger share of the total costs, if those buildings are quite similar.

Estimating Example 1

Your total bid price is $5,000,000. Building A has 15,000 sf and building B has 25,000 sf. The total building area is: 15,000 + 25,000 = 40,000 sf.

Percentage for building A of the total area: 15,000/40,000 × 100% = 37.5%

Percentage for building B of the total area: 25,000/40,000 × 100% = 62.5%

Therefore, the cost of building A is 37.5% × $5,000,000 = $1,875,000; the cost of building B is 62.5% × $5,000,000 = $3,125,000.

Another way is to break out price by the number of functional components (e.g., residential condo units, hotel rooms, or hospital beds).

Estimating Example 2

You total bid price is $6,000,000. Building A has 40 condo units each, and building B has 60 condo units. The total condo units are 40 + 60 = 100 ea.

Percentage for building A of the total condo units: 40/100 × 100% = 40%

Percentage for building B of the total condo units: 60/100 × 100% = 60%

Therefore:

Cost for building A: 40% × $6,000,000 = $2,400,000

Cost for building B: 60% × $6,000,000 = $3,600,000

In reality, cost breakdown can get quite complicated and often it must be submitted at the bid time (sometimes within 24 to 48 hours of bid closing). It is important to communicate these requirements to your subcontractors and suppliers as early as possible, so that they can give you the cost breakdown when submitting their prices.

EXAMPLE COST BREAKDOWN

The following estimating example is simplified exercise, yet involves more than number crunching.

Estimating Example

ABC Contracting
1 Main Street
Anytown, USA 00000
(555) 555-1234

Please provide the cost breakdown for the project (express in words and figures).

Site Construction: _____ ($_____) Dollars

Building Structures: _____ ($_____) Dollars

Building Finishes: _____ ($_____) Dollars

Building Systems: _____ ($_____) Dollars

Other Costs _____ ($_____) Dollars

Total Bid Price: _____ ($_____) Dollars

Cost Breakdown Procedures

1. Study the requirements and make up a list of cost breakdown codes (in this example, the list is "Site Construction", "Building Structures", "Building Finishes", Building Systems" and "Other Costs").

2. In your bid workup sheet, assign one cost breakdown code to each line item. That is, you decide what cost code each item should carry.

3. Summarize your estimate by cost code. This means you add up the costs for items with the same cost code.

Bid Workup Sheet for Cost Breakdown

ABC Contracting
1 Main Street
Anytown, USA 00000
(555) 555-1234

No	Item	QTY	Unit	Item Subtotal	Cost Code
1	Sitework	1	l/s	$385,000	Site Construction
2	Landscaping	1	l/s	$180,000	Site Construction
3	Termite Control	25,000	sf	$2,000	Site Construction
4	Building Concrete	1	l/s	$130,500	Building Structures
5	Building Sidewalk	1,000	sf	$4,000	Building Structures
6	Masonry	1	l/s	$175,000	Building Structures
7	Stone Veneer	1,050	sf	$26,250	Building Structures
8	Foam Insulation	1	l/s	$7,500	Building Structures
9	Structural Steel	1	l/s	$145,000	Building Structures
10	Aluminum Trellis	1	l/s	$30,000	Building Structures
11	Rough Carpentry	1	l/s	$40,000	Building Finishes
12	Window Sills	1	l/s	$4,500	Building Structures
13	Built-up Roofing	1	l/s	$135,000	Building Structures
14	Caulking	1	l/s	$5,600	Building Structures
15	H.M. Doors	3	ea	$2,325	Building Finishes
16	Wood Doors	2	ea	$1,180	Building Finishes
17	Hardware	5	ea	$1,750	Building Finishes
18	Glass and Glazing	1	l/s	$52,000	Building Finishes
19	Windows	1	l/s	$7,560	Building Finishes
20	Stucco	1	l/s	$65,000	Building Finishes
21	Drywall	1	l/s	$115,000	Building Finishes
22	Acoustical Ceilings	1	l/s	$14,530	Building Finishes
23	VCT	1	l/s	$1,250	Building Finishes
24	Painting	1	l/s	$38,000	Building Finishes
25	Toilet Accessories	1	l/s	$8,000	Building Finishes
26	Awnings	1	l/s	$10,000	Building Structures
27	Fire Protection	1	l/s	$28,000	Building Systems
28	Plumbing	1	l/s	$9,000	Building Systems
29	HVAC	1	l/s	$74,580	Building Systems
30	Electrical	1	l/s	$99,000	Building Systems
31	Site Lighting	1	l/s	$13,570	Site Construction

You can see a cost code has been allocated to each item (e.g., "Site Construction" or "Building Structures").

Bid Recap Sheet for Cost Breakdown

ABC Contracting
1 Main Street
Anytown, USA 00000
(555) 555-1234

Description	Rate	Amount	
Labor Total		$27,805	In Bid Workup Sheet
Material Total		$27,450	In Bid Workup Sheet
Subcontractor Total		$1,755,840	In Bid Workup Sheet
Total Item Costs		$1,811,095	In Bid Workup Sheet
Add Material Sales Tax	6%	$1,647	Other Costs
Add Labor Burden	25%	$6,951	Other Costs
Total Direct Costs		**$1,819,693**	
Indirect Costs			
Home Office Overhead	10%	$181,969	Other Costs
Job Overhead		$135,000	Other Costs
Bond		$40,000	Other Costs
Builder's Risk Insurance		$15,000	Other Costs
Financing Interest		$75,000	Other Costs
Permit		$15,000	Other Costs
Impact Fee		$20,000	Other Costs
Total Indirect Costs		**$481,969**	
Total Costs		**$2,301,663**	
Profit	8%	$184,133	Other Costs
Bid Grand Total		**$ 2,485,796**	

Additional items on the bid recap sheet, including sales tax, labor burden, indirect costs, and profit, have been allocated to "Other Costs."

Cost Breakdown Results

To get the cost breakdown result, add the items with the same cost code. For example, add the costs for "Fire Protection," "Plumbing," "HVAC," and "Electrical," because they all carry the same cost code under "Building Systems." The finished cost breakdown is as follows:

Cost Breakdown Summary	
Item	**Amount**
Site Construction	$ 580,570
Building Structures	$ 673,350
Building Finishes	$ 346,595
Building Systems	$ 210,580
Other Costs	$ 674,701
Total Bid	**$ 2,485,796**

Be sure the total costs after the breakdown equal the original total bid price.

CHAPTER 6

POST-BID ESTIMATING

Estimating does not end with faxing a copy of proposal to the owner. Even if you were not successful, your can learn valuable lessons to help improve future chances.

After the job is awarded and construction begins, as an estimator, you can contribute more to the project management team by helping with material purchase and change order pricing.

In this chapter, the following topics will be covered.

- Process of post-bid review
- Common estimating mistakes and tips to reduce them
- Post-bid cost analysis worksheets
- Bid follow-up, bid history tracking, and markup analysis
- Value-engineering ideas
- Update estimate for contract signing
- Turnover meeting for project management
- Estimating for project management
- Change order pricing

POST-BID REVIEW

Immediately after the bid is over, take time to review it the same day. Identify areas for improvement before your next bid.

Create a post-bid document folder and include the following items.

- Copy of bid proposal (proposal forms, cost breakdown, unit price, alternates, bid clarifications, fax confirmation, etc.)
- Copy of bid estimate, including workup sheets and recap sheets
- Drawings and specs
- List of addenda
- Quantity takeoff and pricing worksheets
- Site visit worksheet and photos
- Quotes from suppliers and subcontractors organized by trades

- Bid correspondence with the owner, architect, engineer, and subtrades
- Bond and insurance quotes
- Estimating notes

Take the following steps in your review process.

1. Identify the mistakes you made today and learn from them in the future.
2. Update the subcontractor database to include new competitive subs.
3. Update the cost database for labor, material, and subcontractor unit prices.
4. Perform cost analysis.

COMMON ESTIMATING MISTAKES

- Scope omission: not including required items is perhaps the most serious mistake. Even a "low" mistake is better than "zero."
- Simple math errors: you incorrectly added or subtracted numbers, or used an incorrect formula or conversion factors.
- Measurement errors: you used the wrong scale for reduced-size drawings. For example, when drawings were half size and went unnoticed, then the area you took off was reduced by as much as 75%, even worse than the 50% reduction you assumed.
- Incorrect material prices: you failed to get material price updates from the suppliers and the unit prices you used were outdated.
- Insufficient labor coverage: you were too optimistic about certain items that will take longer to install than you allowed; or crews will not be available when the job starts.
- Underestimated job duration: you will lose profit with extended jobsite overhead, and risk paying for liquated damages if not able to finish on time.
- "Buy-out" price cuts: you intentionally reduced the quotes from subcontractors or suppliers, in hope to increase competitiveness. It will only bring problems down the road, as subs may be unwilling to cut their prices. Even if they do, they will question your business ethics and decide not to bid with you in the future.

TIPS TO REDUCE ESTIMATING MISTAKES

It is almost impossible to achieve a perfect bid. With quality care, however, you can reduce the chance of making mistakes. The following tips may be helpful.

- Be organized and keep your desk clean.
- Check every page of drawings instead of just one or two sheets.
- Read specs at least twice.
- Use estimating checklists.
- Use estimating forms.
- Spend more time on large and expensive cost items.

- Mark drawings when taking off items.
- Prepare more detailed estimates instead of estimating by the square foot.
- Thoroughly figure material and labor for each item instead of applying a combined unit rate or percentage.
- Round up the results in each step of calculation and drop the pennies.
- Have someone else check your estimate and takeoff.
- Use Excel spreadsheets for calculations and check all formulas.
- Compare costs with another similar project on a unit-price basis.
- Always verify site conditions with drawings.
- Ask questions instead of making assumptions.
- Take your time and never rush the estimate even if under stress.

POST-BID COST ANALYSIS

There are at least three types of analyzing work to do after finishing an estimate.

- *Cost per building area:* It makes sense to look at how much per square foot to construct the whole building (e.g., $120/sf for a single-family house). But you should also look at the unit cost per square foot of building area for a specific trade (e.g., $8.50/sf to frame this building).
- *Cost per item quantity:* How much does it cost to supply a prehung wood door? What is the cost to put on plywood sheathing for one square foot of roof area? This kind of information is useful for future budget estimating, when plans and specs are not very clear.
- *Cost as percentage of total:* In relation to total building construction cost, what are the percentages for excavation, foundation, framing, roofing, etc? For example, if foundation work is normally 8% of the total cost, why does it seem to be 12% in this estimate?

Perform cost analysis frequently to accumulate your knowledge in local market prices. Following is an example cost analysis worksheet. Change the format to suit your specific needs. For example, you can further break down the item "finishes" to "drywall," "flooring," "painting," and so forth.

ABC Contracting
1 Main Street
Anytown, USA 00000
(555) 555-1234

Job Name: ABC Restaurant   Estimate No. 332   Estimator: AD

Type of Building: Single story commercial   Location: Any City, Any State

Square Feet: 5116   Functional Units: 252 seats

Proposal Amount: $2,258, 962   Cost per Square Foot: $441.55

Breakdown	Amount	% of Total Bid	$ per sf
General Requirements	$ 518,710	23.0%	$101.39
Sitework	$ 53,189	2.4%	$ 10.40
Concrete	$ 74,016	3.3%	$ 14.47
Masonry	$ 172,534	7.6%	$ 33.72
Metals	$ 106,720	4.7%	$ 20.86
Wood and Plastics	$ 451,137	20.0%	$ 88.18
Thermal and Moisture Protection	$ 127,064	5.6%	$ 24.84
Doors and Windows	$ 139,850	6.2%	$ 27.34
Finishes	$ 210,024	9.3%	$ 41.05
Specialties	$ 23,900	1.1%	$ 4.67
Equipment	$ 0	0.0%	$ 0.00
Furnishings	$ 1,000	0.0%	$ 0.20
Special Construction	$ 0	0.0%	$ 0.00
Conveying System	$ 0	0.0%	$ 0.00
Mechanical	$ 200,819	8.9%	$ 39.25
Electrical	$ 180,000	8.0%	$ 35.18
Total Bid Amount	$2,258,962	100.0%	$441.55

COST MATRIX FOR SIMILAR JOBS

The following is a cost comparison for four buildings of the same type.

ABC Contracting
1 Main Street
Anytown, USA 00000
(555) 555-1234

	Project 1	**Project 2**	**Project 3**	**Today's Bid**
Square Feet	9,500	10,000	8,000	9,000
Total Price	$565,050	$576,000	$515,050	$564,600
Price Per SF	$59	$58	$64	$63
Termite	$ 200	$ 500	$ 400	$ 600
Concrete	$ 120,000	$ 70,000	$ 70,000	$ 110,000
Masonry	INCL	$ 40,000	$ 40,000	$ 36,000
Foam Insulation	$ 1,400	$ 1,600	$ 1,800	$ 2,000
Structural Steel	$ 45,000	$ 40,000	$ 37,000	$ 31,000
Trusses	$ 5,000	$ 4,000	$ 2,500	INCL
Rough Carpentry	$ 35,000	$ 30,000	$ 30,000	$ 25,000
Finish Carpentry	INCL	$ 20,000	$ 18,000	$ 16,000
Roofing	$ 75,000	$ 70,000	$ 80,000	$ 76,000
Insulation	$ 1,500	$ 2,000	INCL	INCL
Caulking	INCL	$ 2,000	$ 1,000	$ 1,500
Doors	$ 4,000	$ 3,000	$ 3,500	$ 4,500
Shutters	$ 10,000	$ 8,000	$ 9,000	$ 8,000
Automatic Doors	$ 9,000	$ 8,000	$ 10,000	$ 10,000
Glass	$ 15,000	$ 16,000	$ 14,000	$ 18,000
Painting	$ 13,000	$ 11,000	$ 15,000	$ 13,000
Drywall	$ 42,000	$ 12,000	$ 12,000	$ 10,000
Stucco	INCL	$ 32,000	$ 25,000	$ 28,000
Ceiling	$ 9,000	$ 4,000	$ 6,000	$ 8,000
VCT and Carpet	INCL	$ 10,000	$ 8,000	$ 10,000
Hard Tile	$ 24,000	$ 13,000	$ 15,000	$ 12,000
Specialties	$950	$ 900	$ 850	$ 1,000
Fire Sprinkler	$ 5,000	$ 8,000	$ 6,000	$ 4,000
HVAC	$ 40,000	$ 60,000	$ 30,000	$ 50,000
Plumbing	$ 20,000	$ 25,000	$ 30,000	$ 20,000
Electrical	$ 90,000	$ 85,000	$ 50,000	**$ 70,000**

BID PROPOSAL FOLLOW-UP

A lot of things can happen after you send in a bid. Waiting for the owner to call back is not necessarily the best way to get a job. Phone the owner a few days after the bid. E-mails can also be used as an alternative in inquiring about the status. Some people are more willing to respond to e-mails than phone calls, as e-mails do not interfere with their work routines. Here are some questions you can ask:

- Has the job been awarded?
- Did you receive all the information you requested?
- Do you have any questions on what you received?
- Is there anything else you need at this point?
- When would be a good time to check back with you?
- What are the important selection criteria?
- How important is price as a selection factor? Are we low?
- What can we do to win your business?
- Could we schedule a meeting with you to discuss the proposal?
- If we did not get the job, who did?
- What was the price difference between our bid and the winning price?

IMPROVING BID-HIT RATIO

Some contractors often bid six or seven jobs before they get one. If you get every job you bid, you are bidding too low. On the other hand, if only one bid turns out to be successful for every 20 or 30 jobs you bid, you are wasting too much time.

Instead of bidding to the same customers over and over, do a little research to make your bid more profitable. Develop a personal bid history and track it monthly, quarterly, and yearly. Track all types of projects you bid on and each customer you bid to: large versus small, hard bid versus negotiated, plans-specs versus design-build, new construction versus renovation, local versus out of town, commercial versus residential versus industrial, public versus private.

As you study these jobs, you will discover certain customers give you more work than others, and you do better with certain kinds of jobs. Then you can focus on the real customers who can give you business. The following is a practical worksheet for your use.

ABC Contracting
1 Main Street
Anytown, USA 00000
(555) 555-1234

Job Name	Job Type	Bid Date	Job Size	Location	Owner	Bid Result

POST-BID MARKUP ANALYSIS

Your bid was $500,000 and the winning bid was $420,000.

In your bid, the total cost (material, labor, and subcontractor) was $400,000 and markup was $100,000 (i.e., 20%).

Possibility 1: Did the competition charge too low on profit?

You have the correct cost ($400,000). Now suppose if the competition was also right, then their markup is only $20,000 (i.e., 4.7% of the total bid of $420,000). It seems they either made a mistake or were too aggressive.

Possibility 2: Did you incorrectly figure the job?

You discover that you made an error and the cost should be $350,000, not $400,000. Now, if the competitors were correct, then their markup is $70,000 (i.e., 16.7% of the total bid of $420,000), which is probably reasonable.

Use the following worksheet for post-bid markup analysis. Note that it may take several jobs to find out the bidding pattern of your competitors.

ABC Contracting
1 Main Street
Anytown, USA 00000
(555) 555-1234

Job Name	Our Total Price	Our Direct Cost	Our Over-head and Profit	Competition's Bid Price	Analysis Conclusion

VALUE ENGINEERING IDEAS

The price you submitted could exceed the owner's original budget. In this case, you could be part of a "value engineering" exercise, which is saving project costs through changing the design. Offering your value engineering ideas can put you in a positive light. The owner may perceive you as a team player and thus be more likely to seal the deal with your company.

The following advice may be helpful.

- Always separate value engineering prices from your base bid. Even if you think you can design buildings better than the architect, do not try to reflect that in the base bid, which should be based on plans and specs as well as the codes.

- Do not waste your time by offering to use cheaper or lower quality material. This is not value engineering, but an insult to your customer. The money savings should be done without sacrificing the building quality (or even as an improvement with less money).

- Keep your overhead when proposing value engineering numbers. For example, if a suggested design change can save the owner $10,000 in total, you should only propose a less amount (say $7,000). Because you are dealing with uncertainties and for the value engineering idea proposed, there may be additional work for you.

Following are common value engineering ideas.

Civil

- Change the building location and orientation on the site map to minimize length of access roads and service connections.
- Work with existing contours to reduce or eliminate retaining walls and minimize volumes of cut and fill.
- Reduce extent of concrete pavers where storm water infiltration is not a priority, and substitute with brushed concrete or asphalt.
- Delete curbs/gutters where possible.
- Use landscaping as a design element to reduce building heat loads.
- Reduce car parking to a minimum.
- Consider smaller, more natural and informal outdoor play areas versus typical regulation size, irrigated, all-weather playing fields.

Architectural and Structural

- Examine alternate structural systems (e.g., wood versus cast-in-place concrete, versus tilt-up concrete, versus concrete block, versus steel joist and decking).
- Set the main floor at level ground to minimize required cut and fill, exterior stairs, foundation walls, retaining walls.
- Minimize roof and floor-to-floor heights.
- Slope the roof to reduce perimeter wall height.
- Limit design variables to increase construction efficiency (e.g., minimize the number of different panel modules for tilt-up concrete structures).
- Consider alternate exterior wall assemblies (e.g., eliminate exterior wall articulation that is not part of an integrated design strategy).
- Eliminate curved walls, especially those with custom-built elements (e.g., radiating beams and odd-sized windows).
- Minimize roof parapet height and decorative form.
- Optimize roof overhangs.
- Examine the amount of roof insulation and roof assembly details to minimize the work required to seal the building.
- Limit the size/extent of entrance canopies and covered walkways.
- Reduce the amount (area) and specifications of glazing on each side.
- Delete sunshades, window eyebrows, and light-shelves.
- Reduce concrete floor slab thickness and use of reinforcing mesh.
- Substitute alternate floor finishes (e.g., 22 ounce level-loop carpet in lieu of other carpet selections; sheet vinyl goods in selected heavy traffic/wear areas in lieu of ceramic tile; integral color concrete versus sheet goods or vinyl composite tile).
- Substitute alternate interior wall finishes (e.g., eliminate decorative vinyl wall coverings or wall paneling in favor of painted gypsum wallboard; limit corner guards and bumper rails to high traffic/wear areas only; delete wainscoting protection; paint masonry wall only when necessary; minimize area and height of ceramic tile wall finish in washrooms).

- Substitute alternate ceiling finishes (e.g., lay-in tile in preference to painted drywall; minimize tile specifications; use 2×4 versus 2×2 tile module).
- Delete decorative gypsum wallboard bulkheads or valences.
- Eliminate side lights and transom lights to doors as well as extensive door glazing. Simply use slot windows in doors.
- Review extent and specifications of millwork against functional needs.

Mechanical

- Review heating and cooling loads and optimize equipment efficiency.
- Size primary mechanical equipment to satisfy load calculations and limit safety factors.
- Compare geothermal system with conventional HVAC system.
- Eliminate the use of "100% outside" air units.
- Review the use of custom size units (i.e., units over 25,000 CFM capacity).
- Use normal electrical-powered chillers in lieu of absorption or gas-operated chillers.
- Use simple air pressure control in lieu of sensor, digital pressure indicator and control.
- Review the use of dehumidification or humidification system.
- Review the use of energy recovery system.
- Consider the alternative materials for piping or ductwork (e.g., fiberglass instead of sheet metal).
- Use flush tank in washrooms in lieu of flush valve.
- Delete perimeter radiation as a secondary heating source.
- Use floating-point valves in lieu of proportional valves.
- Review scope of commissioning.
- Review site grading to see if weeping tile could be reduced.

Electrical

- Substitute load centers for panel boards.
- Combine circuits into common conduit runs.
- Reduce material quality (fixtures, equipment, etc.) to code grade standards.
- Load lighting and plug circuits to the capacity of the circuit breaker protecting the wire.
- Relocate panels to more central locations to cut down the lengths of conduit runs.
- Reduce the size of panel and switchboards to the minimum capacity requirements.
- Eliminate special installation details and instructions wherever possible.
- Use aluminum wire instead of copper feeder wire if allowed.
- Reduce wiring methods to standard code requirements.
- Use instant-start electronic ballasts instead of standard conventional magnetic ballasts.
- Use line-power clocks in lieu of central clock system.
- Delete gypsum wallboard lighting valences in favor of less expensive fixture covers (e.g., sconces).

REVISING ESTIMATE FOR CONTRACT

After being awarded the job, you are informed that the drawings are getting changed "for construction." Before a final contract can be signed with the owner, you must offer a revised price that reflects design changes. Take the following steps.

1. Obtain a complete set of the latest drawings and specs, including civil, landscape, architectural, interior design, structural, and mechanical/electrical.

2. Review documents page by page to see what has changed since the bid. Track and record the differences in detail. Cloud the changes even if the architect forgets to cloud them.

3. Some issues have been discussed with the owner, but the architect has yet to change the drawings. Include the additional costs in your pricing.

4. Check the original list of inclusions, exclusions, and assumptions. See if there are new questions to be added and clarified.

5. Send a new RFI to ask more questions if necessary. Make sure all items will be clarified in writing to your satisfaction.

6. Perform revised quantity takeoff with greater accuracy. If necessary, start a new takeoff and keep it separate from the older version.

7. Get price updates from suppliers and subs. Make sure they have access to the latest design information. Prices may already have changed dramatically since the bid, especially if several months have passed.

8. Get a correct new total number and submit a revised proposal.

TURN-OVER MEETINGS

If you are lucky enough to get a seemingly profitable job, then the next thing you do is to hold a turn-over meeting with your field personnel who will run the daily jobsite operation.

Why have turn-over meetings when everyone involved in the project can simply read a copy of the job estimate for answers? The reason is, by communicating with the estimator who knows the job best, project managers and superintendents are much better informed. They are able to expedite the process of submittals, shop drawings, and purchase orders.

To have a successful meeting, determine who should attend: estimator, project manager, coordinators, superintendents, foreman, and job accountant. External personnel are normally not permitted, as turn-over meetings always involve sensitive information such as contract amount, low subs, and suppliers.

Communication is the key, and the meeting should be more than handing over a package of bid documents. The following example meeting agenda may prove helpful.

1. Have a job overview. Everyone should know the job name and location, building square footage, site acreage, names of owner, architect/engineer, and so forth.

2. Review contract documents, including drawings, specs, addenda, general and special conditions. Examine potential design problems, requests for information, and change orders.

3. Review bid proposal and estimate breakdown. Identify each component that makes up the building. What assumptions did you make in pricing the job? Are they reasonable in today's situation?

4. Review the list of subs and suppliers. Are they reliable to honor their bids? Are they capable of performing the work? Is there going to be a price escalation for some trades?

5. Review the estimate for work performed by your own forces. Did you assume realistic labor productivity in preparing your estimate?

6. Review project schedule and identify the long lead items (e.g., steel, elevator).

7. Review the list of jobsite overhead items. Discuss site conditions and material storage problems (e.g., where to place the tower crane).

8. Evaluate risks. Everyone needs to know the contract amount finally agreed upon with the owner (not necessarily the bid value). List the factors that might pose challenges.

ESTIMATING FOR PROJECT MANAGEMENT

After the job begins, an estimator can still provide continuous support to the project management team. For example, you will need to update the estimate for material purchase, especially when your company self-performs some of the work. The following advice may prove helpful.

- Resolve problems in documents as soon as possible. If there were some problems with the drawings at bid time, then they need to be cleared with the architect. Also make sure your estimate is based on the current documents.

- Update your price. Obtain the latest quotes to decide whether it is more suitable to self-perform or hire subs.

- Consider labor productivity and allow reasonable waste in the estimate. Also make it detailed and clear enough for proper material ordering and delivery.

Job costing codes are especially useful when you self-perform some work. You can have the field foreman report labor and material expenditure by the cost codes. Knowing the collected job cost data and how it compares to the estimated costs is worth more to you than any other data. Such knowledge takes the mystery out of labor units or material buying levels, and enables you to make adjustments for future cost estimates.

CHANGE ORDER PRICING PROCEDURES

1. Review the total change request, and determine if it will involve more than one trade. Make a list of trades to be affected.

2. Review original drawings and specs and compare with revised documents.

3. Notify subcontractors and suppliers about the changes.

4. Make a detailed takeoff of all material and work items.

5. Check the jobsite to verify the actual conditions, make notes and take pictures if necessary.

6. Evaluate the effects of the change order on labor productivity.

7. Adjust the quantity takeoff to reflect the actual situation.

8. Review quotes from subcontractors and suppliers and ensure they truly reflect the changes intended.

9. Complete the pricing and send proposals to the owner for approval.

Tips

- Do not handle a design change without written promise of payment.

- Do not deduct overhead and profit for deductive change orders.

- Remember to request additional time if the change will delay your work.

- Do not forget that you almost always have to pay for cancellations, freight, and restocking charges of materials.

- Inform field workers of the changes so they can stop work in the area affected by the deductive change.

- Keep records of change order requests and associated drawings.

EXCEL SPREADSHEET ESTIMATING

Today estimators are using computers to cut time and effort in the workplace. In many small construction companies, where the owner does the estimating, use of computers frees up additional time for field management and business marketing.

Without purchasing expensive software packages, you can turn simple Excel spreadsheets into powerful estimating tools. This chapter begins with the basics of Excel and covers the whole estimating process including quantity takeoff, pricing, evaluating quotes, adding markups. Regardless of your skill level, after reading this chapter, you will be amazed by what you can do with spreadsheets.

The exercises are self-explanatory. You should have no difficulties advancing through the material. The following notations are used.

- Words in all capital letters refer to certain commands on your keyboard (e.g., press CTRL+C keys to copy a cell).

- Words in bold indicate that something is to be typed into a cell, either a spreadsheet formula or text (e.g., enter =**A2** + **B2** in cell D2).

- Words in italic refer to using your mouse to click a menu or button (e.g., click *File* → *Exit* to close a file).

EXERCISE 1: JOB BASIC INFORMATION

Mission Briefing

This exercise covers the basics of Excel spreadsheet estimating. With building a small house in mind, you start by setting up a simple worksheet, recording basic information of the project. Then you calculate the total square footage of the building using simple spreadsheet formulas. These tasks are simple with the objective to learn spreadsheet fundamentals.

Skill Level: Easy

Suggestions to Readers

- If you have limited knowledge about computers or Excel, it is imperative for you to complete this exercise, step by step, before moving on to the next exercise level.

- If you have previous experience with Excel spreadsheets, then consider skipping this exercise. However, it is always helpful to refresh some skills.

Here is the sketch of a small house for this exercise.

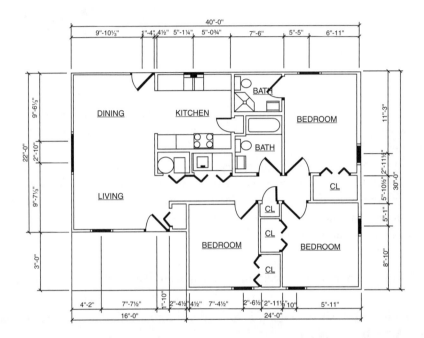

Start Excel

A. Click *Start*

B. Click *All Programs*

C. Click *Microsoft Office*

D. Click *Microsoft Excel w"*

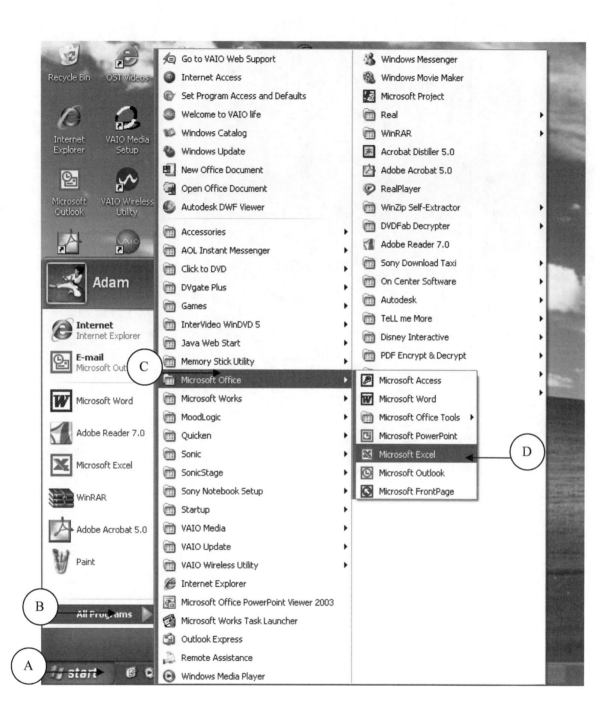

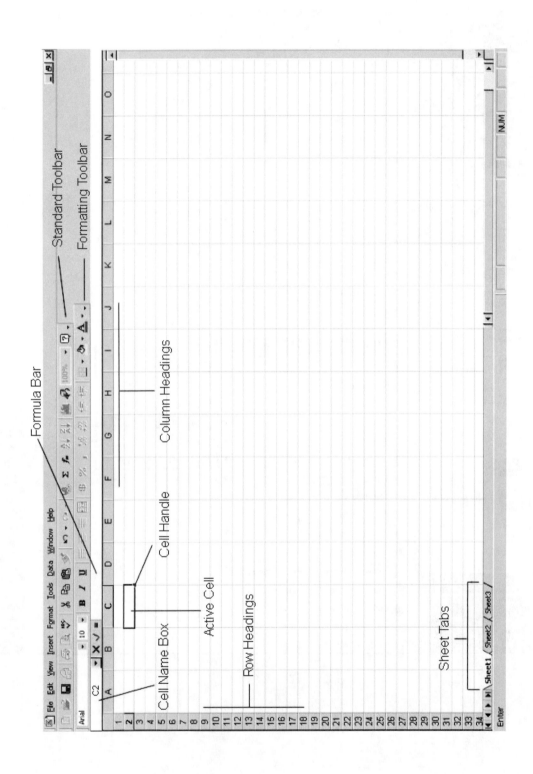

Excel Toolbars

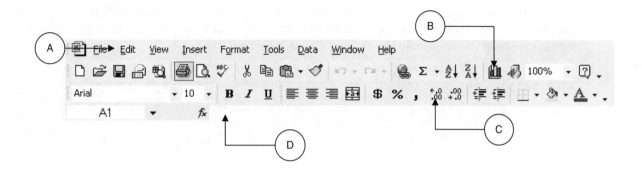

A. *Menu bar*: When you click a menu, pull-down options will show commands available. For example, the "Help" menu provides searchable instructions on how to use Excel.

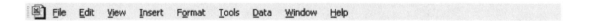

B. *Standard toolbar*: Offers buttons for basic operations (e.g., Create, Open, Save, and Print a worksheet; Cut, Copy, and Paste cell contents; Undo and Redo).

C. *Formatting toolbar*: Offers buttons for cell formats (e.g., fonts, sizes, bold, underline, alignment, merge, dollar sign, percentage, thousand separator, decimals, highlights).

D. *Formula bar*: This is where you enter texts and numbers, or create formulas that perform calculations.

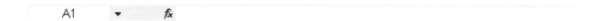

Workbooks, Worksheets, Columns, Rows, and Cells

1. Each Excel file is a our checkbook.

2. In each workbook, there could be many *worksheets*, like each check in your checkbook. By default, Excel puts three worksheets in each new workbook you create (like your bank gives you three blank checks to start). The sheets can be switched by clicking the worksheet tab.

3. In each worksheet, there are *rows* and *columns* of data. Rows are numbered (1, 2, 3, . . .) and columns are lettered (A, B, C, D, . . .). At each intersection of row and column, there is a *cell*, (e.g. "A1", "C3").

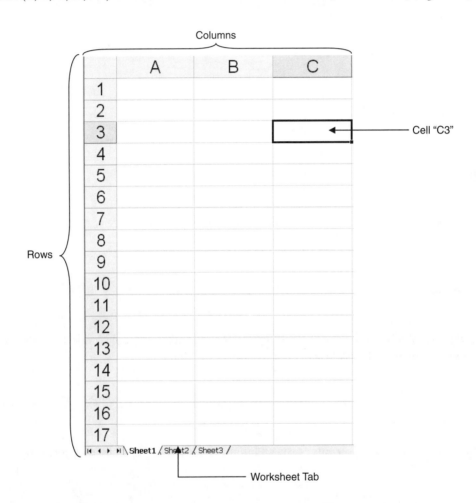

Enter Data, Undo Actions, and Adjust Column Width

1. Use you mouse, select cell A1.

2. Type **Job Name:** and then hit the ENTER key on your keyboard.

3. Cell A2 is now selected. Type **Job Address:** then hit ENTER.

4. Type **Owner's Name:** in cell A3, and **Date:** in cell A4.

5. If you made a mistake, go to menu *Edit → Undo* or just click ↶ button on the toolbar.

6. Column A is a little too narrow for the text entered. To fix this, select column A and double-click the right-side boundary of the column heading "A." (You will see an ↔ indicator appear on the column boundary when you try to do this.)

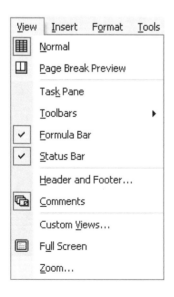

Zoom Display

1. If the screen is too small for you, click *View* on the menu bar, then click *Zoom* on the pull-down menu. Then you can select the magnification level.

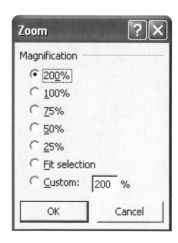

2. Use your mouse and select cell B1. (Alternatively, you can use the four arrow keys on your keyboard to move up, down, left, or right.)

3. Type **Simple Life Residence** in cell B1, **Any Street** in cell B2, and **Adam Ding** in cell B3.

4. Select column B, and double-click the right side boundary of heading B to make the content more visible.

	A	B
1	Job Name:	Simple Life Residence
2	Job Address:	Any Street
3	Owner's Name:	Adam Ding
4	Date:	

Enter Date

1. Select cell B4, type **11/05/2007**, the date November 5, 2007.

2. On the Formatting toolbar, click ▤ button to align the date to the left.

	A	B
1	Job Name:	Simple Life Residence
2	Job Address:	Any Street
3	Owner's Name:	Adam Ding
4	Date:	11/5/2007

3. If you don't like the way the date was displayed, then select cell B4. Right-click your mouse. You will see a drop-down menu. Click *Format Cells*.

4. In the dialog box, select the tab *Number*. Then, under *Category*, select *Date*. To the right side, under *Type*, pick the date/time format you like (scroll down the list to see more).

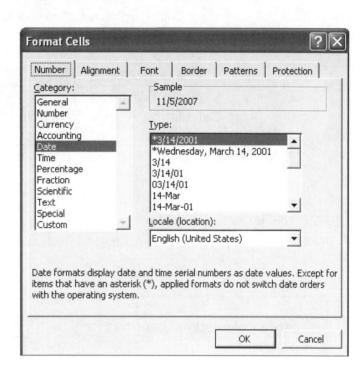

Write Simple Formula

1. Break the house area (refer to the drawing for this exercise) into two rectangles, 22 ft × 16 ft and 30 ft × 24 ft. Then enter the following measurements.

	A	B	C	D	E
6	**Area Calculation**	Length (ft)	Width (ft)	Area	Unit
7	Rect 1	22	16		Sqft
8	Rect 2	30	24		Sqft

2. Select cell A6 and click **B** button on Formatting toolbar to make it bold.

3. Now select cell D7.

4. Move your mouse to the Formula bar and click the blank line behind the formula symbol f_x. You will now see two buttons ✗ ✓ appearing before the symbol f_x.

5. Type = **B7 * C7**. You can find the multiplying operator * on your keyboard.

	A	B	C	D
	SUM ▾ ✗ ✓ f_x =B7*C7			
6	**Area Calculation**	Length (ft)	Width (ft)	Area
7	Rect 1	22	16	=B7*C7
8	Rect 2	30	24	

6. Hit ENTER. The area for rectangle 1 has been calculated.

7. If the value in cells B7 and C7 change, the result will be updated automatically.

	A	B	C	D	E
6	**Area Calculation**	Length (ft)	Width (ft)	Area	Unit
7	Rect 1	22	16	352	Sqft
8	Rect 2	30	24		Sqft

Copy Formula and Auto Sum

1. With cell D7 selected, press CTRL and C keys at the same time on your keyboard (i.e., press CTRL + C).

2. Move to cell D8, press CTRL and V keys at the same time (i.e., CTRL + V).

3. The formula has been copied from cell D7 to cell D8. Hit the ESC key.

	A	B	C	D	E
6	**Area Calculation**	Length (ft)	Width (ft)	Area	Unit
7	Rect 1	22	16	352	Sqft
8	Rect 2	30	24	720	Sqft
9					

4. Select cell A9, type **Total Areas**.
5. Select cell D9. Click AutoSum button Σ on Standard toolbar.

	A	B	C	D	E	F
6	**Area Calculation**	Length (ft)	Width (ft)	Area	Unit	
7	Rect 1	22	16	352	Sqft	
8	Rect 2	30	24	720	Sqft	
9	Total Areas			=SUM(D7:D8)		
10				SUM(**number1**, [number2], ...)		

6. Excel asks you if you want to add up cells D7 and D8.
7. Hit ENTER to confirm. (Only do this if you agree with Excel's suggestion.)
8. Type **Sqft** in cell E9. Select row 9, then click buttons **B** and *I* .

	A	B	C	D	E
6	**Area Calculation**	Length (ft)	Width (ft)	Area	Unit
7	Rect 1	22	16	352	Sqft
8	Rect 2	30	24	720	Sqft
9	*Total Areas*			*1072*	*Sqft*

Set Cell Borders and Check Spelling

1. Select cell A1 through E9. (Use your mouse to drag over the whole area.)
2. Right-click your mouse. From the drop-down menu, pick *Format Cells*.
3. In the dialog box, select the tab *Border*. Then decide the *Style* you want under *Line*. Then click *Outline* and *Inside* buttons under *Presets*. Click *OK*.

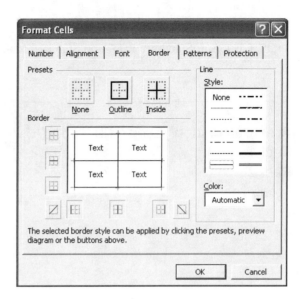

4. With cells still selected, click Spell Check button 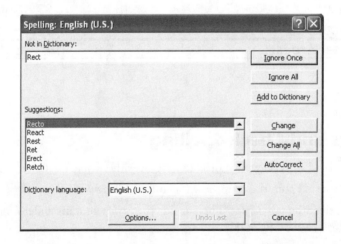 on the Standard toolbar.

5. "Rect" is the short word for rectangle, so click *Ignore All*.

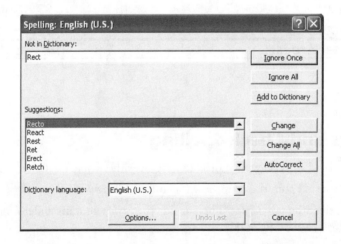

Set Print Area and Print Review

1. With cells A1 through E9 s u *File → Print Area → Set Print Area*. This is to define the content you want to print.

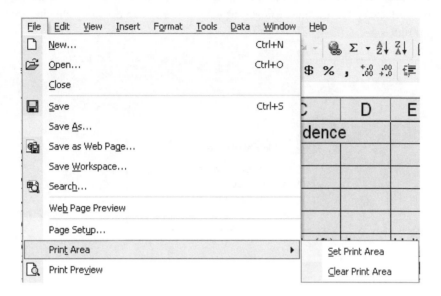

2. Next click menu *File → Print Preview*. This window allows you to see what will be printed before the printer actually prints it.

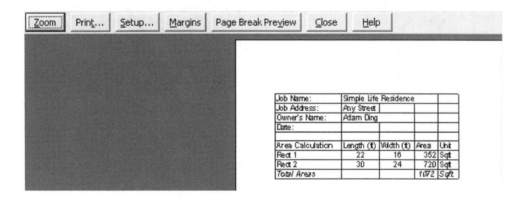

3. If you like what you see, click the *Print* button from this window. Congratulations, you have just printed your first spreadsheet.

Save Workbook and Exit Excel

1. You can click *File → Exit* to close the workbook. Or you can click the small button ✖ on the top right corner of the window to close.

2. If you have not yet saved the workbook during the exercise, you will now be reminded. It is a good habit to save your work every few minutes to prevent data loss. You can do so by clicking this 💾 save button from time to time.

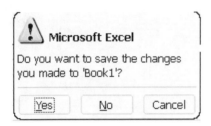

3. Click *Yes*. Then you will be asked where you would like to save it. Pick the location and change the File name **Book1** to something you can remember easily (e.g., **Estimate**). Click *Save*.

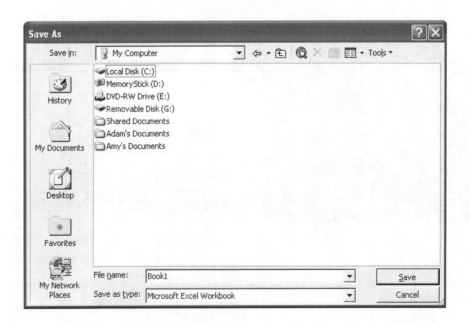

EXERCISE 2: CONSTRUCTION COST SUMMARY

Mission Briefing

This exercise reinforces what you learned in the first exercise. With building the same small house in mind, you set up a cost summary worksheet, listing common cost items in residential construction. Then you enter progress payment (i.e., draws) information and calculate the variance between actual job costs and estimated costs. Some new Excel spreadsheet skills are also introduced.

Skill Level: Easy/Intermediate

Suggestions to Readers

This exercise is good for all readers, regardless of skill level. The first two exercises of this book cover almost everything you need to set up a basic spreadsheet with professional looking results. So it is for your best interests to go through each step of this exercise.

Open Existing Workbook

1. Find the workbook file you completed in the last exercise. Click on it to open.

2. If you cannot find the file, start Excel. Go to *File* menu, click *Open*. Then double-click the directory beside *Look in*. You will see a list of locations.

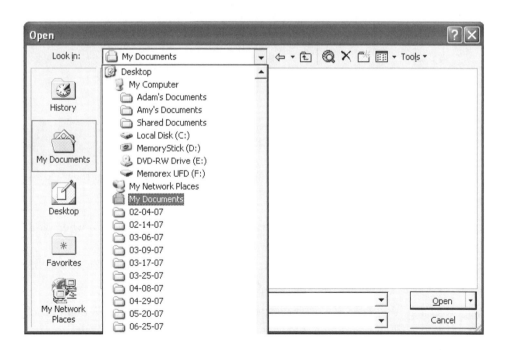

3. After you find the file, double-click on it. This is the screen from the last exercise.

	A	B	C	D	E	F
1	Job Name:	Simple Life Residence				
2	Job Address:	Any Street				
3	Owner's Name:	Adam Ding				
4	Date:					
5						
6	**Area Calculation**	Length (ft)	Width (ft)	Area	Unit	
7	Rect 1	22	16	352	Sqft	
8	Rect 2	30	24	720	Sqft	
9	*Total Areas*			*1072*	*Sqft*	

(Microsoft Excel - ch1)

Rename Worksheet

1. In the Worksheet Tab area, double-click on *Sheet1*, which is the current worksheet. (Or you can right-click on the tab name and select *Rename*.)

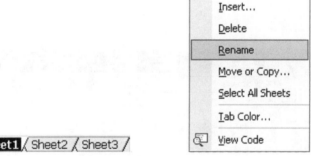

2. With *Sheet1* highlighted, type **Project Info**. Your existing worksheet has been renamed.

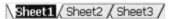

3. Then click on *Sheet2*, and the screen switches to a new blank worksheet. This is where you will enter cost information.

4. Double-click on *Sheet2*. Type **Cost Summary.** The blank worksheet has been renamed, ready for data entry. `\ Project Info \Cost Summary / Sheet3 /`

5. If you want to change the order of worksheets, you can move them around. For example, click on the *Cost Summary* tab, and without releasing it, drag and drop it before the *Project Info* tab.

`\Cost Summary / Project Info / Sheet3 /`

Merge Cells

1. Click on the blank *Cost Summary* worksheet. In cell A1, enter **Simple Life Residence Cost Summary**. It is a little too long to display in one cell.

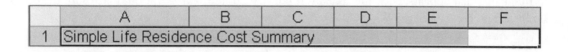

2. Use your mouse to select cells A1 to F1, click Merge button [icon] on toolbar.

3. All cells selected are merged into one. Click Bold button **B** on toolbar.

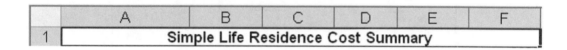

4. Enter a simplified short list of cost items as follows:

	A
1	
2	Jobsite Overhead
3	Excavation
4	Foundation
5	Framing
6	Roof & Siding
7	Insulation
8	Drywall
9	Doors & Windows
10	Cabinets
11	Accessories
12	Painting
13	Floor Coverings
14	Electrical
15	Plumbing and Heating
16	Office Overhead
17	Profit
18	Contingency

Wrap Text

Some cost items, such as "Plumbing and Heating," may be too long to fit into one cell. Instead of expanding your column width or merging cells, you can wrap the text to make the content fit.

1. Select cell A15, right-click your mouse, and pick *Format Cells.*

2. In the dialog that appears, click the *Alignment* tab. Under *Text control* option, make sure the option box *Wrap Text* is checked. Click *OK.*

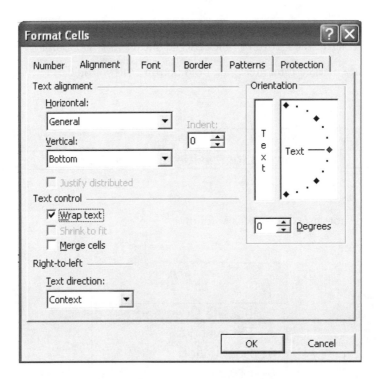

3. Now cell A15 can display the whole text, without making the column too wide.

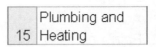

Calculating Cost Variance

On small residential jobs, builders often receive money in three phases: after foundation, after drying-in, and at completion. On larger projects, progress payments are received each month. These payments are typically called "draws."

As the project progresses, you will see some cost items are completed under the budget, while others have an over-run. To figure it out, just compare the total of all draws (actual costs) with estimated costs. The math is as follows:

$$\text{Actual Costs} = \text{Draw 1} + \text{Draw 2} + \text{Draw 3} + \cdots$$

$$\text{Estimated Costs} - \text{Actual Costs} = \text{Cost Variance}$$

If cost variance shows as negative, then you actually spent more money on that item than estimated. In this case, you have an overrun. If cost variance is positive or simply zero, then you completed the item under or within the budget.

The key for budget control is to keep a close look at both the money spent and the work in progress. The two should match. For example, this month you are 20% complete on framing, but how much money is already spent? If 50% or 60% of the money budgeted is already gone, then you may have a problem. Something needs to be done to get the job back on track.

Insert Rows and Columns

1. Click on row 2 to select it. Right-click. Pick *Insert* to make a new row.

2. In the new row, type **Cost Items** in cell A2, **Estimated Costs** in cell B2, **Actual Costs** in cell C2, and **Cost Variance** in cell D2.

	A	B	C	D
1				Simple Life Residence
2	**Cost Items**	Estimated Costs	Actual Costs	Cost Variance
3	Jobsite Overhead			
4	Excavation			
5	Foundation			

3. Select Column C. Go to menu *Insert*, click *Column*.

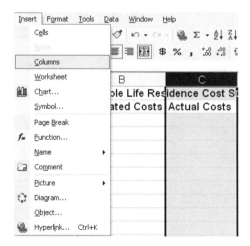

4. Repeat the above step twice. Then you will get three new columns.

5. Type **Draw 1** in cell C2, **Draw 2** in cell D2, and **Draw 3** in cell E2.

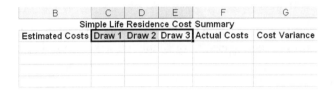

B	C	D	E	F	G
	Simple Life Residence Cost Summary				
Estimated Costs	Draw 1	Draw 2	Draw 3	Actual Costs	Cost Variance

AutoFill Handle

1. Select column A, insert a new column. Type **Ref. #** in new cell A2.

2. Type **'001** in cell A3, and **'002** in cell A4. To do so, make sure you put the symbol of single quote (i.e., apostrophe) before numbers 001 and 002.

3. Select both cell A3 and cell A4. Move your mouse to the lower-right corner of the selection. You will see a black cross. This is your fill handle.

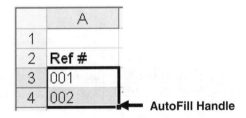

AutoFill Handle

4. Without releasing your mouse, drag it down across the cells you want to fill. When you come to the end of the column, release the mouse button.

5. Without further typing, each cost item in the list now has a unique reference number, or cost code. Auto-Fill Handle is a quick way to enter sequential data.

	A	B
1		
2	**Ref #**	**Cost Items**
3	001	Jobsite Overhead
4	002	Excavation
5	003	Foundation
6	004	Framing
7	005	Roof & Siding
8	006	Insulation
9	007	Drywall
10	008	Doors & Windows
11	009	Cabinets
12	010	Accessories
13	011	Painting
14	012	Floor Coverings
15	013	Electrical
16	014	Plumbing and Heating
17	015	Office Overhead
18	016	Profit
19	017	Contingency

Freeze Panes

Freezing panes means keeping the upper rows always visible when you scroll downward of the screen, or keeping the first few left columns always visible when you scroll to the right side of the screen.

1. Select cell C3. In this case, row 3 is immediately below the title row, and column C is the start of cost information for each item.

2. Go to menu *Window*, click *Freeze Panes.* You will see two lines are now drawn to the above and left of cell C3.

	A	B	C	D
1				
2	**Ref #**	**Cost Items**	**Estimated Costs**	**Draw 1**
3	001	Jobsite Overhead		
4	002	Excavation		
5	003	Foundation		
6	004	Framing		
7	005	Roof & Siding		

Window menu: New Window, Arrange..., Hide, Unhide..., Split, Freeze Panes

3. Now try to scroll down. You will see rows 1 and 2 are always visible. Scroll to the right, columns A and B are always visible.

	A	B	C	D
1				
2	Ref #	Cost Items	Estimated Costs	Draw 1
10	008	Doors & Windows		
11	009	Cabinets		
12	010	Accessories		
13	011	Painting		
14	012	Floor Coverings		
15	013	Electrical		
16	014	Plumbing and Heating		
17	015	Office Overhead		
18	016	Profit		
19	017	Contingency		

	A	B	G	H
1			le Life Residence Cost Summa	
2	Ref #	Cost Items	Actual Costs	Cost Variance
3	001	Jobsite Overhead		
4	002	Excavation		
5	003	Foundation		
6	004	Framing		
7	005	Roof & Siding		
8	006	Insulation		
9	007	Drywall		

Paste Format

Enter the following numbers in the worksheet for this exercise.

	A	B	C	D	E	F
1				Simple Life Residence C		
2	Ref #	Cost Items	Estimated Costs	Draw 1	Draw 2	Draw 3
3	001	Jobsite Overhead	7000	4000	2500	500
4	002	Excavation	1500	800	0	500
5	003	Foundation	6000	7500	0	0
6	004	Framing	20000	0	18000	2500

1. Select cell C3, which has a number (i.e., 7,000) to be shown as a dollar value.
2. Click the dollar symbol **$** on the Format toolbar. (Or you can right-click, select *Format Cells* → *Number* tab → *Category* → *Currency*.)
3. If you wish, click ⊧≣ button on the Format toolbar twice to take out the decimals.

C
Estimated Costs
$ 7,000.00

C
Estimated Costs
$ 7,000.0

C
Estimated Costs
$ 7,000

4. Now you need to apply the same currency format to the rest of worksheet.
5. With cell C3 still selected, click the brush symbol 🖌 on the Standard toolbar to copy its format (not the cell itself). Then your mouse turns into a combination of cross and brush.
6. Drag your mouse across cells C3 through F6 to change them into dollars.

	A	B	C	D	E	F
1				Simple Life Residence Cos		
2	Ref #	Cost Items	Estimated Costs	Draw 1	Draw 2	Draw 3
3	001	Jobsite Overhead	$ 7,000	$4,000	$ 2,500	$ 500
4	002	Excavation	$ 1,500	$ 800	$ -	$ 500
5	003	Foundation	$ 6,000	$7,500	$ -	$ -
6	004	Framing	$ 20,000	$ -	$ 18,000	$ 2,500

Enter Simple Formula

1. Select cell G3. Click AutoSum button Σ on the Standard toolbar.

2. Cell G3 turns into =**Sum(C3:F3)**. Apparently, this is not correct. Actual costs should be the total of three draws (i.e., cells D3, E3, F3), but excluding the value of *Estimated Costs* (cell C3).

	C	D	E	F	G
		Simple Life Residence Cost Summary			
	Estimated Costs	Draw 1	Draw 2	Draw 3	Actual Costs
	$ 7,000	$4,000	$ 2,500	$ 500	=SUM(C3:F3)
	$ 1,500	$ 800	$ -	$ 500	SUM(number1, [nu
	$ 6,000	$7,500	$ -	$ -	

3. Do NOT hit ENTER. Instead, grab your mouse and select cells D3 through F3. You will see the formula in cell G3 automatically change into = **Sum(D3:F3)** when you do this.

	C	D	E	F	G
		Simple Life Residence Cost Summary			
	Estimated Costs	Draw 1	Draw 2	Draw 3	Actual Costs
	$ 7,000	$4,000	$ 2,500	$ 500	=SUM(D3:F3)
	$ 1,500	$ 800	$ -	$ 500	SUM(number1, [nu
	$ 6,000	$7,500	$ -	$ -	

4. Now hit ENTER. Cell G3 becomes $7,000, which is the total of three draws.

	A	B	C	D	E	F	G
1				Simple Life Residence Cost Summary			
2	Ref #	Cost Items	Estimated Costs	Draw 1	Draw 2	Draw 3	Actual Costs
3	001	Jobsite Overhead	$ 7,000	$4,000	$ 2,500	$ 500	$ 7,000
4	002	Excavation	$ 1,500	$ 800	$ -	$ 500	
5	003	Foundation	$ 6,000	$7,500	$ -	$ -	
6	004	Framing	$ 20,000	$ -	$ 18,000	$2,500	

Note that the AutoSum button should be used with caution. What Excel wants to add is not necessarily correct. Alternatively, you can manually enter = **SUM(D3:F3)**.

More on Entering Simple Formula

This exercise is to see if the item "Jobsite Overhead" is under or over budget.

1. Select cell H3, which is the cell under the heading "Cost Variance."
2. Type an equal sign =. This is to tell Excel you are entering a formula.

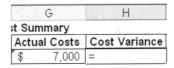

	G	H
	st Summary	
	Actual Costs	Cost Variance
	$ 7,000	=

3. Grab your mouse and select the cell under "Estimated Costs" (i.e., cell C3).

	C	D	E	F	G	H
	Simple Life Residence Cost Summary					
	Estimated Costs	Draw 1	Draw 2	Draw 3	Actual Costs	Cost Variance
	$ 7,000	$4,000	$ 2,500	$ 500	$ 7,000	=C3

4. Type a minus sign - after that.

	G	H
	Summary	
	Actual Costs	Cost Variance
	1000	=C3-

5. Grab your mouse again and select the cell under "Actual Costs" (i.e., cell G3).

	C	D	E	F	G	H
	Simple Life Residence Cost Summary					
	Estimated Costs	Draw 1	Draw 2	Draw 3	Actual Costs	Cost Variance
	$ 7,000	$4,000	$ 2,500	$ 500	$ 7,000	=C3-G3

6.   Hit ENTER. "Cost Variance" is calculated as zero. The item is on budget.

Alternatively, you can directly type a formula **=C3-G3. But** with mouse clicking, you never have to think about cell names like C3 or G3. You just look at the headings (such as "Estimated Costs" or "Actual Costs") and click the appropriate cells to include them in the formula. This clicking method of entering formula is more visual and foolproof.

Copy Formula

The formulas in cells G3 and H3 must be copied to the rest of the worksheet, but instead of keying **Ctrl + C** and **Ctrl + V** dozens of times, you can use AutoFill Handler again to copy formulas.

1.   Select both cells G3 and H3. In the lower-right corner of the selection, find the black cross or AutoFill handle.

2.   Without releasing your mouse, drag the handle down until you reach row 6.

3.   It seems the item "Excavation" is on budget, and items "Foundation" and "Framing" are having overruns.

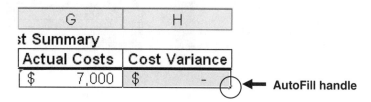

	A	B	C	D	E	F	G	H
1				Simple Life Residence Cost Summary				
2	Ref #	Cost Items	Estimated Costs	Draw 1	Draw 2	Draw 3	Actual Costs	Cost Variance
3	001	Jobsite Overhead	$ 7,000	$4,000	$ 2,500	$ {◊}	$ 7,000	$ -
4	002	Excavation	$ 1,500	$ 800	$ -	$ 500	$ 1,300	$ 200
5	003	Foundation	$ 6,000	$7,500	$ -	$ -	$ 7,500	$ (1,500)
6	004	Framing	$ 20,000	$ -	$18,000	$2,500	$ 20,500	$ (500)

Page Setup

1. Click menu *File → Print Review*. Or simply click button 🔍 on toolbar.

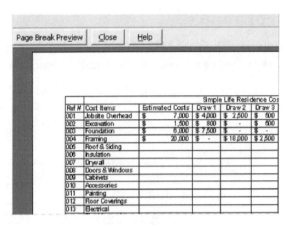

2. Apparently the page is not wide enough to display all content. Without leaving this window, click *Setup* on the toolbar above.
3. In the dialog window that appears, under *Page* tab, change *Orientation* from *Portrait* to *Landscape*. Click *OK*. (Other options, such as print scale, paper size, and print quality, are also available for adjustment.)

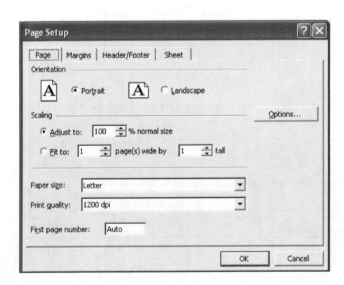

Add Header and Footer

1. In the same *Page Setup* window, click the tab *Header/Footer*.

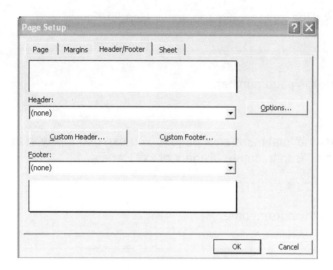

2. Then click *Custom Header*.
3. In the window that appeared, click the space under *Center section*. Type **Cost Summary Sheet**. Click *OK*.

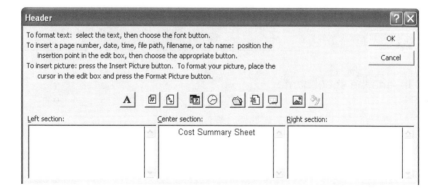

4. Then click *Custom Footer*.
5. In the similar window that appeared, click the space under *Right section*.
6. Then click the symbol 🔢 to insert page number. Click *OK*.
7. Click *OK* to dismiss the dialog window. Click *Print* to print the worksheet.

Following is a list of icons in the page setup dialog window and their meanings.

A	Change the font for selected text in Header/Footer.
#	Inserts the page number.
	Inserts the total number of pages. You can use this with the "Insert Page Number" icon to create "Page x of xxx Pages".
	Inserts the current date.
	Inserts the current time.
	Inserts the path and file name.
	Inserts the file name.
	Inserts the sheet name.
	Inserts a picture.
	Format the picture inserted

EXERCISE 3: QUANTITY TAKEOFF
Mission Briefing

In this exercise, you will work on:

- Calculating Lengths, Areas and Volumes
- Converting Measurement Units
- Estimating Excavation
- Estimating Concrete and Formwork
- Estimating Wood Framing

Skill Level: Intermediate

Suggestions to Readers

It is recommended for all readers to master this exercise as the foundation block of spreadsheet estimating. This exercise is perhaps the most important part of this chapter. It provides intensive training on how to write simple formulas for calculations (i.e., add, subtract, multiply, and divide cells). Predefined worksheet functions are also discussed. Completed worksheets in this exercise can be used repeatedly as quantity takeoff templates.

Converting Lengths

For easier estimating, it is recommended to convert all lengths, including fractions of an inch, to linear feet in decimal format.

Worksheet Setup

	A	B	C	D	E	F
1	Feet	Inch	Fraction		Result	
2	5	4	3/8		5.36	Feet

1. Enter row 1 texts as shown.
2. Enter **5** for cell A2 and **4** for cell B2. Do not include feet and inch symbols (i.e., single or double quote marks).
3. Select cell C2. Right-click and pick *Format Cells*. (Or, simply press CTRL and number 1 keys at the same time.)

4. In the dialog box that follows, make sure *Number* tab is active. Select *Fraction* under *Category*. Then pick *as eighths* under *Type*. (Normally 1/8 inch is a precision level good enough for estimating.)

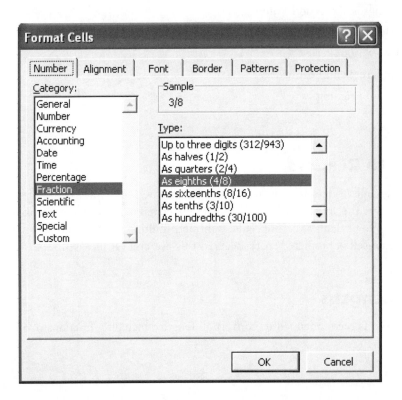

5. Click *OK*. In cell C2, try to put in a fraction (e.g., **3/8**). Again do not include the inch symbol after the fraction entered.

6. Now in cell E2, type a formula = **A2+B2/12+C2/12** (i.e., feet + inch/12 + fraction of an inch/12). Note at the beginning you should type an equal sign. This is to tell Excel you are entering a formula in this cell.

7. Hit ENTER. The result is calculated. You now have a length converter.

8. Next time, simply change the numbers in cell A2, B2, or C2, and the result will be automatically recalculated.

Calculating Areas

	A	B	C	D	E	F
1	Shape	Length (Ft)	Width (Ft)		Area (SF)	Area In SY
2						
3	Rectangle	15	20		300	33.3
4						
5	Triangle	15	20		150	16.7
6		Diameter (Ft)				
7	Circle	15			177	19.6
8		Top Width (Ft)	Bottom Width (Ft)	Height (Ft)		
9	Trapezoid	15	20	25	438	48.6

1. Enter all texts as shown.
2. Enter only numbers for length, width, diameter, and so forth, in columns B, C, and D. Do not include foot or inch symbols (single or double quote marks).
3. In cell E3, type **= B3*C3** (i.e., Length × Width for Rectangle).
4. In cell E5, type **= 1/2*B5*C5** (i.e., 1/2 × Length × Width for Triangle).
5. In cell E7, type **= 0.7854*B7*B7** (i.e., 0.7854 × Diameter² for Circle).
6. In cell E9, type **= 1/2*(B9+C9)*D9** (i.e., 1/2 × [Top Width + Bottom Width] × Height for Trapezoid).
7. In cell F3, type **=E3/9** to covert square feet to square yards.
8. With cell F3 selected, press CTRL + C on the keyboard to copy its formula.
9. Hold down SHIFT key, select cells F5, F7, and F9 one by one.
10. Press CTRL + V on the keyboard to paste the copied formula.
11. Press ENTER key. Now you have an area calculator that can be reused.

Calculating Volumes

Most volume calculations are simply Length × Width × Height, though some shapes are a bit more complicated. With Excel, formulas are built into the spreadsheets and even the cubic feet are automatically converted to cubic yards.

Worksheet Setup

	A	B	C	D	E	F	G
1	Shape	Length (Ft)	Width (Ft)	Height (Ft)		Volume (CF)	In CY
2							
3	Prism	100	3	1.5		450	16.7
4							
5		Diameter (Ft)		Height (Ft)		Volume (CF)	In CY
6	Cylinder	2		5		15.7	0.6

1. Enter all texts as shown. Enter only numbers in columns B, C, and D. Do not include foot or inch symbols (i.e., single or double quote marks).
2. In cell F3, type **= B3*C3*D3** (i.e., Length × Width × Height).
3. In cell F6, type **= 0.7854*B6*B6*D6** (i.e., 0.7854 × Diameter^2 × Height).
4. In cell G3, type **= F3/27** to convert cubic feet to cubic yards.
5. With cell G3 selected, press CTRL + C on the keyboard to copy the formula.
6. Select cell G6, press CTRL + V on the keyboard to paste the formula.
7. Press ESC key. Now you have a volume calculator that can be reused.

Estimating Basement Excavation

Worksheet Setup

	A	B	C
1	*Assuming Excavation Slope of 1:1*		
2			
3	Building Length	60	Feet
4	Building Width	35	Feet
5	Excavation Depth	10	Feet
6			
7	**Estimate Results**		
8	Building Area	2100	Sq. Ft
9	Building Perimeter	190	Sq. Ft
10	Excavation Volume (net)	1130	Cubic Yards
11	Backfill Volume (net)	352	Cubic Yards

1. Cell B8 calculates building area: $= B3*B4$ (i.e., Building Length $\times$ Building Width).
2. Cell B9 calculates building perimeter: $= (B3+B4)*2$ (i.e., [Length + Width] $\times$ 2).
3. Cell B10 figures excavation volume in cubic yard based on excavation slope.

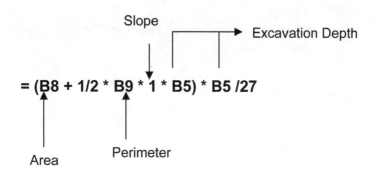

4. Cell B11 figures backfill volume in cubic yard: $= B10-B8*B5/27$ (i.e., excavation volume minus the basement volume).

Estimating Strip Footing

Worksheet Setup

	A	B	C	D	E	F	G
	Description	Footing Length (LF)	Footing Width (LF)	Footing Depth (LF)		Formwork (SFCA)	Concrete Volume (CY)
2	SF1	40	2	1		84	3.0
3	SF2	22	2.5	1.5		73.5	3.1
4	SF2	24	2.5	1.5		79.5	3.3
5	SF3	24	1.5	0.67		34.2	0.9
6	SF2	30	2.5	1.5		97.5	4.2
7							
8	Total					369	14

1. Enter lengths and dimensions for different footing sections in columns B, C, and D.
2. In cell F2, type $= (B2+C2)*2*D2$ (i.e., Formwork = [Footing Length + Footing Width] $\times$ 2 $\times$ Footing Height).

3. In cell G2, type **= B2*C2*D2/27** (i.e., Concrete = Footing Length × Footing Width × Footing Height, then convert cubic feet to cubic yard).

4. Select both cells F2 and G2, in the lower-right corner of the selection, find a black cross or AutoFill handle **+**.

5. Drag the fill handle down to copy the formula down to the next four rows.

6. In the bottom cells F8 and G8, click the AutoSum button **Σ** on the toolbar to summarize the total formwork area and concrete volume.

7. To reuse this sheet as a takeoff template, simply insert a new row. Enter the new dimensions and then copy the formula down for automatic calculations.

Estimating Pad Footings

Worksheet Setup

	A	B	C	D	E	F	G	H
1	Description	Number of Pads (EA)	Pad Length (LF)	Pad Width (LF)	Pad Depth (LF)		Formwork (SFCA)	Concrete Volume (CY)
2	F1	2	2	2	1.5		24	0.4
3	F2	4	1.5	1.5	1		24	0.3
4								
5	Total						48	0.8
6								

The setup is similar to continuous footing in the previous example.
Formulas are as follows:

1. In cell G2, **= B2*(C2+D2)*2*E2** (i.e., Formwork = Number of Pads × [Pad Length + Pad Width] × 2 × Pad Depth).

2. In cell H2, **= B2*C2*D2*E2/27** (i.e., Concrete = Number of Pads × Pad Length × Pad Width × Pad Depth, then convert cubic feet to cubic yards).

The completed worksheet can also be reused as a takeoff template.

Estimating Concrete Walls
Worksheet Setup

	A	B	C	D	E	F	G	H
1	Description	Wall Length (LF)	Wall Thickness (LF)	Wall Height (LF)		Wall Area (SF)	Formwork (SFCA)	Concrete Volume (CY)
2	Grid A	40	1	10		400	820	14.8
3	Grid 1	30	0.67	9		270	552	6.7
4	Grid C	40	0.83	8		320	653	9.8
5	Grid 5	30	0.67	9		270	552	6.7
6								
7	Total					1260	2577	38.1

Formula Explanations

- In cell F2, = **B2*D2** (i.e., Wall Area = Wall Length × Wall Height).
- In cell G2, = **(B2+C2)*2*D2** (i.e., Formwork = [Wall Length + Wall Thickness] × 2 × Wall Height).
- In cell H2, = **B2*C2*D2/27** (i.e., Concrete = Wall Length × Wall Thickness × Wall Height, then convert cubic feet to cubic yards).

Note that all dimensions are entered in decimal format for easy calculations (e.g., 0.67 ft for 8 inch wall thickness, 0.83 ft for 10 inches, 1 ft for 12 inches).

Estimating Slabs and Driveways
Worksheet Setup

	A	B	C	D	E	F	G	H
1	Description	Slab Length (LF)	Slab Width (LF)	Slab Thickness (LF)		Slab Area (SF)	Formwork (SFCA)	Concrete Volume (CY)
2	Garage	30	24	0.33		720	36	8.8
3	Driveway	20	24	0.33		480	29	5.9
4								
5	Total					1200	65	14.7

Formula Explanations

- In cell F2, = **B2*C2** (i.e., Slab Area = Slab Length × Slab Width).
- In cell G2, = **(B2+C2)*2*D2** (i.e., Edge Formwork = [Slab Length + Slab Width] × 2 × Slab Thickness, if the slab needs to be formed all the way around).
- In cell H2, = **B2*C2*D2/27** (i.e., Concrete = Slab Length × Slab Width × Slab Thickness, then convert cubic feet to cubic yards).

Note that slab thickness is entered in decimal format for easy calculations (0.33 ft for 4 inches in this example).

Estimating Dimensional Lumber

Worksheet Setup

	A	B	C	D	E	F	G	H	I
1	Description	Grade	Quantity (EA)	Length (LF)	Section Thickness (Inch)	Section Width (Inch)		Dimension	Board Feet
2	Roof Blocking	P.T.	2	150	2	4		2 x 4	200
3	Floor Beam		3	10	2	6		2 x 6	30
4	Posts		4	9	4	4		4 x 4	48

Formula Explanations

- In cell H2, **= E2&"×"&F2**, the ampersand symbol (& without dot) is used to connect the texts. Here two **&** are used to connect section thickness and section width, using an "×" in between. Then the lumber dimension will be automatically displayed each time, based on your input.

- In cell I2, **= C2*D2*E2*F2/12** (i.e., Board Feet = Quantity of Lumber Pieces × Piece Length × Section Thickness × Section Width/12).

You can improve the sheet by indicating grade as well in the result.

	A	B	C	D	E	F	G	H	I
1	Description	Grade	Quantity (EA)	Length (LF)	Section Thickness (Inch)	Section Width (Inch)		Dimension	Board Feet
2	Roof Blocking	P.T.	2	150	2	4		P.T. 2 x 4	200
3	Floor Beam		3	10	2	6		2 x 6	30
4	Posts		4	9	4	4		4 x 4	48

Formula for cell H2 is: **= B2&E2&"x"&F2** (i.e., Grade + Dimensions).

Estimating Plywood Sheets

Worksheet Setup

	A	B	C	D	E	F	G	H	I	J
1	Description	Thickness (Inch)	Grade	Quantity (EA)	Length (LF)	Width (LF)		Dimension	Sheets	Round Up
2	Floor	5/8"		1	35	22		5/8"	24.06	25
3	Ext Wall	1/2"	CDX	1	100	9		1/2"CDX	28.13	29
4	Canopy	3/4"	FT	2	30	5		3/4"FT	9.38	10

Formula Explanations

- In cell H2, = **B2&C2** (i.e., Thickness + Grade for Plywood).
- In cell I2, = **D2*E2*F2/32** (i.e., Pieces of Plywood × Coverage Length × Coverage Width/32, if the plywood sheet is in standard size of 4 ft × 8 ft).
- In cell J2, = **ROUNDUP(I2,0)** (i.e., round up the result to whole plywood sheets).

The predefined worksheet function ROUNDUP () with zero inside is used to round up the calculation result to the nearest whole number. For more related information on this useful worksheet function, press F1 key and search help for ROUNDUP.

EXERCISE 4: ADVANCED QUANTITY TAKEOFF

Mission Briefing

In this exercise, you will work on:

- Calculating Triangle Hypotenuse
- Estimating Masonry
- Estimating Drywall
- Estimating Reinforcing Steel
- Estimating Sloped Roof Rafter
- Estimating Interior Floor Finishes

Skill Level: Advanced

Suggestions to Readers

For those who have prior experience in working with spreadsheets, this exercise will be an exciting eye-opener. It introduces useful Excel tips that may be new to you and will revolutionize the way you look at spreadsheets (if used properly). So if you are wondering what Excel can do beyond adding numbers, you will enjoy the material.

Calculating Triangle Hypotenuse

The longest side of a right triangle is called the hypotenuse (side c in the diagram). You can find its length if the two other sides of the triangle (sides a and b) are known.

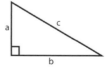

Formula for length of hypotenuse: $c = \sqrt{a^2 + b^2}$.
The idea is simple:

1. Take squares on each of the two shorter sides.
2. Add them up.
3. Take the square root of the sum to find out hypotenuse length.

Worksheet Setup

	A	B	C
1	**Triangle Sides**	**Length**	**Unit**
2	Side One	5	FT
3	Side Two	12	FT
4			
5	Hypotenuse	13	FT

Formula Explanations

Formula in cell B5: **= SQRT (B2^2+B3^2)**

- The symbol ^ is called caret, and is located on your keyboard above the number 6 key. To type it, press SHIFT + 6.

- The caret is an exponentiation math operator. "B2^2" means raise cell B2 value to power 2, or take squares on it. In the above example, B2^2=5^2=25 and B3^2=12^2=144.

- Worksheet function SQRT () is used to take a square root on a value. In the above example, 25 + 144 = 169, thus SQRT (169) = 13.

- Try to change one side to 3 and the other side to 4. You will see the hypotenuse automatically recalculates to 5.

- If you do not want to use the caret, your formula in cell B5 would have been: **= SQRT(B2*B2+B3*B3)**. It works in the same way.

Estimating Masonry

In estimating masonry, the essential idea is to find out how many blocks or bricks are needed for the project, and everything else will be based on that amount. In Excel worksheet, the quantity of blocks/bricks can be stored in a cell whose location never changes. Then you can give this cell a name. Other calculations such as mortar and sand can use the given name without referring to any cell addresses.

Worksheet Setup

	A	B	C
1	**Item**	**QTY**	**Unit**
2	**Input**		
3	Block Wall Length	100	FT
4	Block Wall Height	3	FT
5			
6	**Estimate Results**		
7	Wall Area	300	Sqft
8	Number of Blocks	338	EA
9			
10	Mortar	10	Bag
11	Sand	39	CF
12	Cell fill	84	CF

1. Enter all texts as shown. Type **100** in cell B3 and **3** in cell B4.
2. Enter formula = **B3*B4** in cell B7.
3. Enter formula = **B7*1.125** in cell B8 (i.e., 1.125 blocks per square feet of wall area).
4. With cell B8 selected, press CTRL + F3 keys from the keyboard.
5. In the dialog that follows, give cell B8 a simple name. Type **Block** (instead of a long name, **Number_of_ Blocks**). Click *OK*.

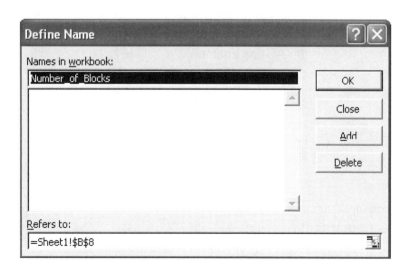

6. Now the name "Block", or the value of cell B8, is ready for use.

7. In cell B10, enter formula **= 3*Block/100** for mortar (i.e., three bags of mortar per 100 blocks).

8. In cell B11, enter formula **= 11.6*Block/100** for sand (i.e., 11.6 cf of sand per 100 blocks, or 1 cy of sand per seven bags of mortar).

9. In cell B12, enter formula **= Block*0.25** for cell fill grout (i.e., 0.25 cf of grout per block, excluding bond beams).

Formula Notes

To define a cell name, you do not have to press CTRL + F3 keys and then go through that dialog window. Alternatively, just type the name **Block** in the Name box at the left end of formula bar, like this:

| Block | ▼ | *fx* |

. It will work in the same way.

Estimating Drywall

Like masonry, the main concern of drywall takeoff is the wall area to be covered (or sheets of drywall). Everything else will follow this number. The same idea of cell naming can be applied here. Store the number of gypsum boards in a cell. Then give it a name that is easy to remember. All calculations thereafter will just refer to that name.

Worksheet Setup

	A	B	C
1	**Item**	**QTY**	**Unit**
2	**Input**		
3	Wall length	15	FT
4	Wall height	9	FT
5	Sides to be covered	2	EA
6	Layers on each side	2	EA
7			
8	**Estimate Results**		
9	Wall coverage area	540	SF
10	Gypsum boards	17	EA
11			
12	Ready mix	77	LBS
13	Tape	262	LF
14	Joint compound	49	LBS
15	Nails	2.7	LBS
16	Screws	680	EA

1. Enter all texts and input numbers as shown.

2. Enter formula **= B3*B4*B5*B6** in cell B9 for wall coverage area.

3. Enter formula **= ROUNDUP(B9/32,0)** in cell B10 (for 4 ft × 8 ft sheets).

4. With cell B10 selected, go to the Name box at the left end of the formula bar (where it shows "B10"). Substitute it with **DrywallBoard,** by simply typing the word. Remember to hit ENTER to make the name effective.

DrywallBoard ▼	𝑓𝑥

5. Now the name "DrywallBoard," or the value of cell B10, is ready for use.
6. In cell B12, enter formula = **DrywallBoard*4.5** (i.e., 4.5 lb of ready mix per sheet of drywall).
7. In cell B13, enter formula = **DrywallBoard*15.4** (i.e., 15.4 ft of tape per sheet of drywall).
8. In cell B14, enter formula = **DrywallBoard*2.9** (i.e., 2.9 lb of joint compound per sheet of drywall).
9. In cell B15, enter formula = **DrywallBoard*0.16** (i.e., 0.16 lb of nails per sheet of drywall).
10. In cell B16, enter formula = **DrywallBoard*40** (i.e., 40 ea screws per sheet of drywall).

Formula Notes

1. To name cells, you can also press CRTL + F3 keys and then go through the dialog window.
2. It is important to use the same *exact* spelling for names throughout. For example, you defined a cell to be "DrywallBoard," but then in the formulas that followed, you refer it as "Drywall Board." With that extra space between two words, it will not work. Excel will give you a "Name?" error message.

Estimating Reinforcing Steel

Worksheet Setup

A. Rebar Weight Data Table

	A	B
1	Data Table	
2	Bar Size	Unit Weight (Lbs /Ft)
3	#2	0.167
4	#3	0.376
5	#4	0.668
6	#5	1.043

1. Set up a data table for bar sizes and their unit weights. For simplicity, just use bars #2 to #5. Please do NOT put spaces between "#" symbol and numbers.
2. Select cell range A2:B6 (to do so, use your mouse to drag from the top left corner of cell A2 to the lower right corner of cell B6).
3. Press CTRL + F3 key. In the dialog window, give this data table a meaning name (e.g., change "Bar_Size" to "**RebarWeight**").

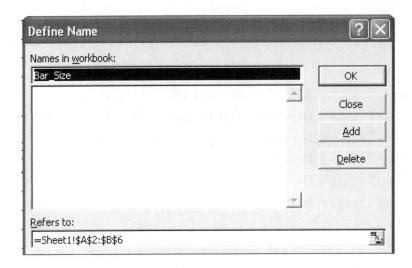

4. Click *OK*. Now the whole rebar weight table can simply be referred to as "RebarWeight." You can see that name shown on the formula bar like this: RebarWeight ▼ *fx* (click on the arrow if you cannot see it).

B. Rebar Weight Estimate Worksheet

	A	B	C	D	E	F
8	**Estimate Sheet**					
9	Description	Number of Bars (EA)	Bar Length (LF)	Bar Size	Unit Weight (Lbs/Ft)	Bar Weight (Lbs)
10	Footing	2	180	#5	1.043	375.5

Go to a few rows below the rebar weight table, start an estimate sheet like the one shown above.

This estimate sheet uses LOOKUP worksheet function, so you do not have to manually look up the rebar unit weight each time. Instead, simply specify the pieces of rebar, its length, and strength, Excel will look up the weight value and automatically calculate the result.

Formula in cell E10: = **LOOKUP(D10,RebarWeight)**

- "D10" is *what* you want to look for ("#5" rebar size).

- "RebarWeight" is *where* you want to look for (the name of rebar weight data table, which actually refers to cell range A2:B6).

- LOOKUP () function generates what you want: the value from the second column of rebar weight table (i.e., 1.043 lb/ft for #5 bar you looked for).

Formula in cell F10: = **B10*C10*E10** (i.e., Bar Weight = Number of Bars × Bar Length × Rebar Unit Weight). Note that this is net quantity and does not consider lapping, splicing and waste.

This estimate sheet, like most other worksheets in this chapter, can be used as a takeoff template. To do so, keep entering new data in the next row, and simply copy the formula down to make it work.

Estimating Roof Rafter

Excel can solve sloped roof problems by establishing a worksheet that can be reused. You no longer have to look up the conversion factors every time. You simply specify the slope, and Excel will take care of the rest.

Worksheet Setup

A. Roof Slope Conversion Factor Table

1. Enter the above table as shown. Select cells A2 through B12.
2. Go to menu *Insert → Name → Define*.

	A	B
1	**Roof Slope**	**Conversion Factor**
2	2 in 12	1.014
3	3 in 12	1.031
4	4 in 12	1.054
5	5 in 12	1.083
6	6 in 12	1.118
7	7 in 12	1.158
8	8 in 12	1.202
9	9 in 12	1.250
10	10 in 12	1.302
11	11 in 12	1.357
12	12 in 12	1.413

3. In the highlighted name box (where it shows "_2_in_12"), type **SlopeTable**. Click *OK*. Now the whole data table can be referred by a name as "SlopeTable."

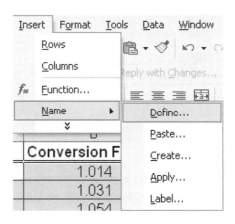

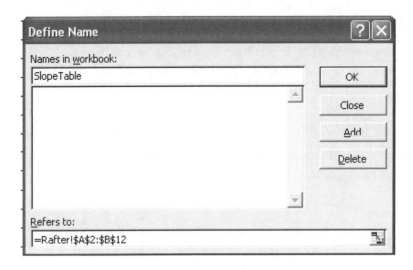

4. Select cell range A2 through A12 (without conversion factors in column B).

5. Repeat the above process and give the range A2:A12 a name **RoofSlope**. Now you should be able to see both names from the drop-down list on the formula bar.

B. Roof Rafter Estimate Worksheet

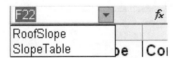

6. Start a new estimate sheet right below the data table created.

	A	B	C	
14	**Estimate Sheet**			
15	**Description**	**Run Length (LF)**	**Roof Slope**	**Conv**
16	Arbor Framing	15	5 in 12	
17			5 in 12	
18			6 in 12	
19			7 in 12	
			8 in 12	
20			9 in 12	
21			10 in 12	
			11 in 12	
22			12 in 12	

7. Enter text and numbers in columns A and B as shown.

8. Select column C by clicking on the column heading.

9. Go to menu *Data*, click *Validation*.

10. In the dialog window that appears, select *List* under *Allow*.

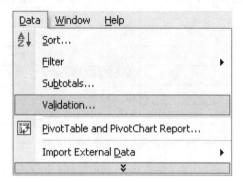

11. Under *Source*, type **= RoofSlope**. Remember to include the equal sign.

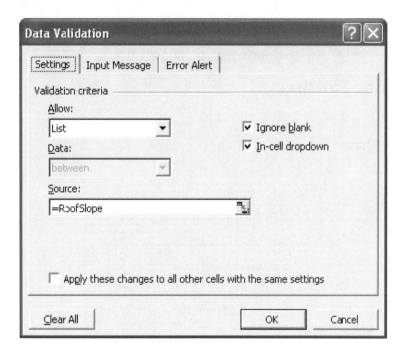

12. Click *OK*. Now go to any cell in column C like cell C16, you will find a downward arrow besides it. When clicking on the arrow, you will have a list of roof slopes to choose from. This is also what you are allowed to enter (if you enter something else, Excel will give you an error message).

13. Now you are ready to put formulas in columns D and E to complete the calculation.

	A	B	C	D	E
14	**Estimate Sheet**				
15	**Description**	**Run Length (LF)**	**Roof Slope**	**Conversion Factor**	**Rafter Length (LF**
16	Arbor Framing	15	5 in 12 ▼	1.083	16.245
17			5 in 12 ▲		
			6 in 12		
18			7 in 12		
			8 in 12		
19			9 in 12		
20			10 in 12		
21			11 in 12		
			12 in 12 ▼		
22					

In the above worksheet, data validation has been used to limit the data entry (roof slopes), so LOOKUP function can find *exactly* what you are looking for, with no mistakes.

Formula in cell D16: = **LOOKUP(C16,SlopeTable)**

- "C16" is what you want to look for (e.g., "5 in 12" roof slope).
- "SlopeTable" is where you want to look for (the name of conversion factor table, or cell range A2:B12).
- LOOKUP ( ) gives the search result: the value from the 2nd column of slope table (i.e., 1.083 for slope 5:12).

Formula in cell E16: = **B16*D16** (i.e., Sloped Roof Rafter Length = Run Length × Slope Factor).

Estimating Floor Finishes

Worksheet Setup

A. List of Floor Finish Types

	A	B
1	**Floor Finish**	**Description**
2	CER	Ceramic Tile
3	RESIL	Resilient flooring
4	CRPT	Carpet

1. Start a table listing all floor finish types. Use abbreviations such as CER, CRPT if necessary. Give explanations of what these abbreviations mean.
2. Select cell range A2:A4 (without the heading A1).
3. Go the formula bar where it says "A2." Type **FloorType**. Hit ENTER to make the name effective.
4. Now the cell range A2:A4 has been given the name "FloorType."

FloorType	▼	*fx*

B. Floor Finish Schedule

5. Start another table right below (on the same worksheet), listing the name and square footage for each room, based on your measurements. Leave floor finish information blank.

	A	B	C	D
6	**Room**	**Area**	**Unit**	**Floor**
7	Entry	130	SF	
8	Family Rm	118	SF	
9	Dining	180	SF	
10	Living	152	SF	
11	Corridor	122	SF	
12	Closet	42	SF	
13	Laundry	36	SF	
14	Storage	31	SF	
15	Bedroom #1	145	SF	
16	In-suite Bath	56	SF	
17	Bedroom #2	140	SF	
18	Bathroom	35	SF	
19	Kitchen	95	SF	

6. Select cell D7:D19. Go to menu *Data → Validation*.

7. Pick *List* under *Allow*. Type = **FloorType** under *Source*.

8. Click *OK*.

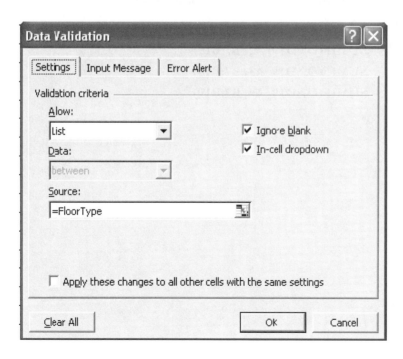

	A	B	C	D
6	Room	Area	Unit	Floor
7	Entry	130	SF	RESIL
8	Family Rm	148	SF	CER
				RESIL
9	Dining	180	SF	CRPT
10	Living	152	SF	CRPT
11	Corridor	122	SF	CRPT
12	Closet	42	SF	CRPT
13	Laundry	36	SF	RESIL
14	Storage	31	SF	RESIL
15	Bedroom #1	145	SF	CRPT
16	In-suite Bath	56	SF	CER
17	Bedroom #2	140	SF	CRPT
18	Bathroom	35	SF	CER
19	Kitchen	95	SF	RESIL

9. Now enter floor finish for each room by picking from the drop-down list in each cell, based on job specifications.

C. Floor Finish Area Estimate Summary

10. Now go back to the first table, and summarize floor finish areas.

Formula Explanations

Formula in cell C2: = **SUMIF(D7:D19, A2, B7:B19)**

Formula in cell C3: = **SUMIF(D7:D19, A3, B7:B19)**

Formula in cell C4: = **SUMIF(D7:D19, A4, B7:B19)**

	A	B	C	D
1	Floor Finish	Description	Total	Unit
2	CER	Ceramic Tile	91	SF
3	RESIL	Resilient flooring	472	SF
4	CRPT	Carpet	749	SF
5				
6	Room	Area	Unit	Floor
7	Entry	130	SF	RESIL
8	Family Rm	148	SF	CRPT
9	Dining	180	SF	RESIL
10	Living	152	SF	CRPT
11	Corridor	122	SF	CRPT
12	Closet	42	SF	CRPT
13	Laundry	36	SF	RESIL
14	Storage	31	SF	RESIL
15	Bedroom #1	145	SF	CRPT
16	In-suite Bath	56	SF	CER
17	Bedroom #2	140	SF	CRPT
18	Bathroom	35	SF	CER
19	Kitchen	95	SF	RESIL

- "D17:D19" is what you want to evaluate (floor finish in each room).
- "A2" is a test condition you specify (e.g., Is the room finish ceramic tile?).
- "B7:B19" is what you want to add up when the test condition is met (e.g., add the room area only if it is finished with ceramic tile).
- SUMIF function adds up all room areas that meet the test condition

In this worksheet, SUMIF worksheet function is used to do selective adding, whereas features such as data validation and naming cell range make this task even easier. So whenever a room changes flooring (e.g., from carpet to vinyl), simply pick a different finish type for that room; and the total floor finish quantities will be automatically recalculated.

EXERCISE 5: ESTIMATING MULTIFAMILY HOUSING
Mission Briefing

This exercise focuses on estimating for multifamily housing (i.e. apartments, townhomes or hotels), whether they are low-rise wood-framed or high-rise complicated concrete structures. With the help of one simple spreadsheet function called SUMPRODUCT(), you will estimate the following with ease:

- Total gross floor area (including floor area, perimeter and height at each floor level), total area of living spaces and underground parking
- Total number of residential suites and the suite count at each floor level
- The count of bathrooms based on suite types
- Total interior area of residential suites
- Total common living area (e.g., elevator, lobby, stairway, corridors)

- Area of suite balcony decks and patios
- Area of interior floor finishes
- Area of different exterior claddings
- Area of interior partitions, including suite demising walls, corridor walls, and suite interior walls
- Count of suite doors listed by sizes and types

Skill Level: Intermediate/Advanced (All readers are encouraged to try the material due to its importance.)

Measuring Gross Floor Area

In this example low-rise condominium building, parking is underground. Concrete slab on grade is at the lowest level, with concrete columns supporting suspended concrete slabs and slab bands from above. Wood-framed structures are from the main floor level and going up to the roof.

	A	B	C	D
1	**Floor Level**	**Area**	**Perimeter**	**Flr Height**
2		Sqft	Ft	Ft
3	Parking	20,000	700	11
4	Level 1	16,000	750	10
5	Level 2	15,000	760	10
6	Level 3	15,000	760	10
7	Level 4	7,000	450	9
8				
9	**Total GFA**	**73,000**	**Sqft**	
10				
11	**Breakdown**			
12	Parking Area	20,000	Sqft	
13	Living Spaces	53,000	Sqft	

Formula in cell B9, = **SUM(B3:B7)**, the sum of areas from all floors

Formula in cell B12, = **B3**, the area of underground parking only

Formula in cell B13, = **B9-B12** or = **SUM(B4:B7)**, the area of four upper floors for living spaces.

Counting Number of Suites

There are three prototypes of condo units ("A", "B", "C") in this building. It is helpful to know the number of units per floor level as well as for each type.

	A	B	C	D	E
1	Units	"A"	"B"	"C"	Subtotal
2	Type	1-bed/1-bath	2-bed/1-bath	2-bed/2-bath	
3	Level 1	6	5	3	14
4	Level 2	5	5	4	14
5	Level 3	5	5	4	14
6	Level 4	5		2	7
7	Subtotal	21	15	13	
8					
9	Total Units	49			

Formula in cell B7, = **SUM(B3:B6),** 21 of unit "A"

Formula in cell C7, = **SUM(C3:C6),** 15 of unit "B"

Formula in cell D7, = **SUM(D3:D6),** 13 of unit "C"

Formula in cell E3, = **SUM(B3:D3),** 14 units on Level 1

Formula in cell E4, = **SUM(B4:D4),** 14 units on Level 2

Formula in cell E5, = **SUM(B5:D5),** 14 units on Level 3

Formula in cell E6, = **SUM(B6:D6),** 7 units on Level 4

Formula in cell B9, = **SUM(B7:D7)** or = **SUM(E3:E6),** 49 units total.

Counting Total Number of Suite Bathrooms

Traditional Estimating Method

To figure the total number of bathrooms, find out how many bathrooms in each type of condo unit, then multiply that by the number of condo units, add them up to reach a total. (Note that this exercise does not consider any public bathrooms in common areas.)

Estimating Math

Total QTY of items = Σ (Item QTY in each condo unit $\times$ Number of condo units)

	A	B	C	D
1	Units	"A"	"B"	"C"
2	Type	1-bed/1-bath	2-bed/1-bath	2-bed/2-bath
3	Total Number of Units	21	15	13
4	Bathroom in Each Unit	1	1	2
5				
6	Bathrooms Subtotal	21	15	26
7				
8	Total Bathrooms	62	EA	

Formula in cell B6, = **B3*B4**, 21 of unit "A" and they are one-bath units

Formula in cell C6, = **C3*C4**, 15 of unit "B" and they are one-bath units

Formula in cell D6, = **D3*D4**, 13 of unit "C" and they are two-bath units

Formula in cell D8, = **SUM(B6:D6)**, 62 bathrooms total

Using SUMPRODUCT Worksheet Function

	A	B	C	D
1	Units	"A"	"B"	"C"
2	Type	1-bed/1-bath	2-bed/1-bath	2-bed/2-bath
3	Total Number of Units	21	15	13
4	Bathroom in Each Unit	1	1	2
5				
6	Total Bathrooms	62	EA	

Formula in cell B6: = **SUMPRODUCT(B3:D3*B4:D4)**

SUMPRODUCT worksheet function multiplies the values of corresponding cells in two lists, and then adds up the products. In this example, the two lists are as follows:

- B3:D3, the total nuvmber of units "A", "B," and "C"
- B4:D4, the number of bathrooms in each unit "A", "B," and "C"

What SUMPRODUCT function does is take the quantity of each unit type (21 of unit "A" in cell B3), find the corresponding bathroom (1 bathroom for unit "A" in cell B4), and multiply the two ($21 \times 1 = 21$). It does the same thing for the rest of two-unit types (15×1 for unit "B" and 13×2 for unit "C"). Finally it adds up the multiplying results ($21 + 15 + 26 = 62$). You see Excel uses the same math but does everything in one single step!

SUMPRODUCT is one of the most significant Excel features for estimating multifamily housing. Complicated projects can have many floors and lots of suite prototypes. SUMPRODUCT will make your estimating much easier.

Estimating Suite Floor Area

To estimate total interior floor area in condo units (often referred as "sellable area"), normally you will measure the square footage in each unit type, multiply the number of units in each type, and then add them up. For example:

"A": 800 sf $\times$ 21 = 16,800 sf

"B": 1,050 sf $\times$ 15 = 15,750 sf

"C": 1,100 sf $\times$ 13 = 14,300 sf

Add them up: 16,800 + 15,750 + 14,300 = 46,850 sf

Using SUMPRODUCT worksheet function, however, you can skip many steps.

	A	B	C	D
1	Units	"A"	"B"	"C"
2	Total Number of Units	21	15	13
3	Square Footage in Each Unit	800	1050	1100
4				
5	Total Unit Area	46850	Sqft	

Formula in cell B5: = **SUMPRODUCT(B2:D2*B3:D3)**

The two lists SUMPRODUCT function used are as follows:

- B2:D2, the total number of units "A", "B," and "C"
- B3:D3, the square footage in each unit "A," "B," and "C"

It's that simple!

Estimating Common Living Area

Each floor level has common living areas, such as electrical closets, stairways, and elevator shafts. It is helpful to know out of the total 53,000 sf of living space, how many square feet are interior condo areas and how many square feet are common areas. It also helps to know how the numbers break down floor by floor.

	A	B	C	D
1	Units	"A"	"B"	"C"
2	Level 1	6	5	3
3	Level 2	5	5	4
4	Level 3	5	5	4
5	Level 4	5	0	2
6	Total Number of Units	21	15	13
7				
8	Unit	"A"	"B"	"C"
9	Area in Each Unit	800	1050	1100
10				
11	Living Space Level	Living Area	Unit Area	Common Area
12	Level 1 Area	16,000	13,350	2,650
13	Level 2 Area	15,000	13,650	1,350
14	Level 3 Area	15,000	13,650	1,350
15	Level 4 Area	7,000	6,200	800
16	Total	53,000	46,850	6,150

To determine unit areas per floor, the formulas are as follows:

In cell C12: = **SUMPRODUCT(B2:D2*B9:D9)**, Level 1 Units × Their Areas

In cell C13: = **SUMPRODUCT(B3:D3*B9:D9)**, Level 2 Units × Their Areas

In cell C14: = **SUMPRODUCT(B4:D4*B9:D9)**, Level 3 Units × Their Areas

In cell C15: = **SUMPRODUCT(B5:D5*B9:D9)**, Level 4 Units × Their Areas

Common area is simply deducting unit area from total area on each floor (e.g., formula in cell D12 is: = **B12-C12**).

Estimating Suite Decks

In this apartment building, upper floor units have decks covered by vinyl waterproof membrane and main floor ones are patios covered by landscape pavers. This exercise calculates the vinyl deck areas for upper floor units.

	A	B	C	D
1	Upper Floor	"A"	"B"	"C"
2	Level 2	5	5	4
3	Level 3	5	5	4
4	Level 4	5		2
5	**Total Number of Units**	**15**	**10**	**10**
6				
7				
8	**Unit Deck**	"A"	"B"	"C"
9	**Deck Area (SF) in Each Unit**	100	150	200
10				
11	**Upper Level**	**Deck Area**		
12	Level 2 Deck Area	2,050	SF	
13	Level 3 Deck Area	2,050	SF	
14	Level 4 Deck Area	900	SF	
15				
16	**Total Deck Area**	**5,000**	**SF**	

To figure out deck areas per floor, formulas are as follows:

In cell B12: = **SUMPRODUCT(B2:D2*B9:D9)**, Level 2 Units × Their Decks

In cell B13: = **SUMPRODUCT(B3:D3*B9:D9)**, Level 3 Units × Their Decks

In cell B14: = **SUMPRODUCT(B4:D4*B9:D9)**, Level 4 Units × Their Decks

You may have noticed giving cell range B9:D9 a name (e.g., "decks") will simplify these formulas. If a name was defined, the formula for cell B12 would have been = **SUMPRODUCT(B2:D2*decks)**.

Estimating Suite Floor Finishes

In each condo unit, sheet vinyl is specified for kitchen and bath area, while carpet is for the remainder (bedrooms, living rooms, den, entry, closets, etc.). You can measure the vinyl and carpet areas in each unit type, and then use SUMPRODUCT function to reach total floor finish areas.

	A	B	C	D
1	**Units**	**"A"**	**"B"**	**"C"**
2	Total Number of Units (EA)	21	15	13
3	Vinyl (SF) in Each Unit	140	170	180
4	Carpet (SF) in Each Unit	660	880	920
5				
6	**Subtotal**			
7	Total Vinyl in Units	7,830	**SF**	
8	Total Carpets in Units	39,020	**SF**	

Formula in cell B7: = **SUMPRODUCT(B2:D2*B3:D3)**

That is: Number of Units (B2:D2) × Vinyl Area in Each Unit Type (B3:D3).

Formula in cell B7: = **SUMPRODUCT(B2:D2*B4:D4)**

That is: Number of Units (B2:D2) × Carpet Area in Each Unit Type (B4:D4).

Estimating Building Total Floor Finishes

Now if floor finish for common living area is added (see previous exercises on how to calculate common area), you can figure out the total floor finish areas for the entire building.

	A	B	C
7	Total Vinyl in Units	7,830	**SF**
8	Total Carpets in Units	39,020	**SF**
9	Carpets in Common Area	6,150	**SF**
10			
11	**Flooring Summary**		
12	Vinyl flooring	7,830	SF
13	Carpets	45,170	SF
14	Unfinished parking	20,000	SF
15	**Total**	**73,000**	**SF**

Formula in cell B12: = **B7**, vinyl in suites

Formula in cell B13: = **B8+B9**, carpet in suites and in common areas

Formula in cell B14: = **SUM(B12:B14)**, total floor area for the building

Total floor finish area (including sealed concrete in parking level) should equal to the gross floor area of the building. This is also a double check of your calculations.

Estimating Exterior Walls

To estimate the area for different types of exterior wall, builders often go through exterior elevations and take off the area for each type. Then add up all four sides of the buildings for each type of wall. In the above worksheet, formula in cell F2 is = **SUM(B2:E2)**. This is to add up the vinyl siding areas from all sides.

	A	B	C	D	E	F	G
1	Ext Wall	South	East	North	West	Total	
2	Vinyl siding	3,500	4,850	3,240	3,790	15,380	SF
3	Brick	2,110	2,030	980	2,530	7,650	SF
4	Stone	850	350	630	500	2,330	SF
5					Total Area	25,360	SF

But often due to the zigzag of building footprint, some walls will not show on any exterior elevations, thus being missed. Here's a quick way to check this.

	A	B	C	D	E
7	Floor Level	Perimeter	Flr Height	Wall Area	
8		Ft	Ft	SF	
9	Level 1	750	10	7,500	
10	Level 2	760	10	7,600	
11	Level 3	760	10	7,600	
12	Level 4	450	9	4,050	
13	Total			26,750	SF

Simply take the perimeter of each floor level above grade, multiply it by floor height to get the exposed wall area. In the above worksheet, formula in cell D9 is = **B9*C9**. You can see the total wall area by this method is larger than from the elevations alone. You may have missed some walls that were only shown on floor plans.

Estimating Interior Partitions

Demising walls are dividing walls that separate one condo unit from another. To take off demising walls, simply measure the total wall length on each floor, then multiply that by wall height. (Note that framing may go above the ceiling.) Add the areas of demising walls from all floor levels.

	A	B	C	D	E
1	Demising Wall	Wall Length	Wall Height	Wall Area	Unit
2	Level 1	265	9	2,385	SF
3	Level 2	265	9	2,385	SF
4	Level 3	265	9	2,385	SF
5	Level 4	150	8	1,200	SF
6			Total	8,355	SF

Formula in cell D2: = **B2*C2**. The next three cells in column D are similar.

Formula in cell D6: = **SUM(D2:D5)**

Corridor walls go along the entrances of all condo units and separate them from the common corridor on each floor level. The takeoff of corridor walls is similar to that of demising walls.

	A	B	C	D	E
8	Corridor Wall	Wall Length	Wall Height	Wall Area	
9	Level 1	350	9	3,150	SF
10	Level 2	350	9	3,150	SF
11	Level 3	350	9	3,150	SF
12	Level 4	200	8	1,600	SF
13			Total	11,050	SF

Unit walls are inside the condo units, separating one room from another. They are also taken off floor by floor. On each floor, you will first measure the wall length in a certain condo type, and then multiply the number of condos on this floor level. Add each condo type to reach a total unit wall length on this floor.

	A	B	C	D	E
1	Units	"A"	"B"	"C"	
2	Level 1	6	5	3	
3	Level 2	5	5	4	
4	Level 3	5	5	4	
5	Level 4	5	0	2	
6	Total Number of Units	21	15	13	
7					
8	Unit Wall Length	130	170	180	
9					
10	Unit Walls	Wall Length	Wall Height	Wall Area	
11	Level 1	2,170	9	19,530	SF
12	Level 2	2,220	9	19,980	SF
13	Level 3	2,220	9	19,980	SF
14	Level 4	1,010	8	8,080	SF
15			Total	67,570	SF

Formula in cell B11: = **SUMPRODUCT(B8:D8*B2:D2)** for unit walls on Level 1. Note the formula used cell range B2:D2, the number of condos in Level 1 only. It did not use cell range B6:D6, the total number of condo units. Cell range B8:D8 is the total interior wall length in each condo type (130 ft of walls in condo "A," 170 ft in "B," and 180 ft in "C").

Similarly in cell B12, the formula is = **SUMPRODUCT(B8:D8*B3:D3)** for Level 2.

In cell B13, the formula is = **SUMPRODUCT(B8:D8*B4:D4)** for Level 3.

In cell B14, the formula is = **SUMPRODUCT(B8:D8*B5:D5)** for Level 4.

Counting Suite Doors

For the conclusion of this exercise, you will count doors in condo units (doors in common areas are omitted for clarity).

	A	B	C	D	E	F	G
1	Units	"A"	"B"	"C"			
2	Total # of Units	21	15	13			
3							
4	Unit Doors	"A"	"B"	"C"	Total Doors		Type
5	Unit Entry Door	1	1	1	49	EA	Wood SC 3068
6	Bedroom Door	1	2	2	77	EA	Wood HC 2868
7	Bathroom Door	1	1	2	62	EA	Wood HC 2668
8	Storage Rm Door	0	1	1	28	EA	Wood HC 2868
9	Closet Door	1	1	1	49	EA	24" -2 Panel
10	Closet Door	1	2	2	77	EA	48" -2 Panel
11	Patio/Balcony Door	1	1	1	49	EA	Steel insulated 2868

Take the following steps.

1. Name cell range B2:D2 as "**Condo,**" so that it can be referred to easily.

2. Count the number of different types of doors in unit "A," "B," and "C," and enter numbers in cells from B5 to D11.

3. Enter formula in cell E5, = **SUMPRODUCT(Condo*B5:D5).**

4. Copy the formula down to cell E11.

5. Add some notes in column G about the door type and dimension.

EXERCISE 6: PRICING

Mission Briefing

In this exercise, you will work on the following.

- Pricing materials
  - Pricing concrete materials
  - Pricing lumber materials
  - Evaluating joist/truss suppliers
- Pricing labor
  - Pricing concrete labor
  - Pricing framing labor
- Combined labor and material pricing
- Evaluating subtrade quotes

  - Evaluating concrete contractors
  - Evaluating wood framers

Skill Level: Easy/Intermediate

Suggestions to Readers

This is a good lesson to refresh formula skills and learn important concepts such as relative and absolute cell references. Some useful worksheet functions will also be introduced.

Pricing Concrete Material

Waste for Concrete Volumes

Formula in cell F2: = **C2*E2** (i.e., Waste = Concrete Volume × Waste Factor)

Formula in cell G2: = **C2+F2** (i.e., Adjusted QTY = Original QTY + Waste)

Formulas in row 3 to row 7 are similar. (Drag down both cell F2 and cell G2 to copy the formulas over.).

	A	B	C	D	E	F	G	H
1	Item	Strength	QTY	Unit	Waste Factor	Waste	Adjusted QTY	Unit
2	Ftg/Pads	3500 PSI	140	CY	10%	14	154	CY
3	Fdn Wall	3500 PSI	250	CY	5%	13	263	CY
4	Interior SOG	3000 PSI	190	CY	10%	19	209	CY
5	Suspended Slab	4000 PSI	420	CY	5%	21	441	CY
6	Column/Stairs	4000 PSI	28	CY	5%	1	29	CY
7	Slab Topping	3000 PSI	145	CY	5%	7	152	CY

Concrete Material Accessories

	A	B	C	D
1	Item	Strength	Adjusted QTY	Unit
2	Ftg/Pads	3500 PSI	154	CY
3	Fdn Wall	3500 PSI	263	CY
4	Interior SOG	3000 PSI	209	CY
5	Suspended Slab	4000 PSI	441	CY
6	Column/Stairs	4000 PSI	29	CY
7	Slab Topping	3000 PSI	152	CY
8				
9	Accesssories	Where It Applies	QTY	Unit
10	Plasticizers	Everywhere	1,248	CY
11	Hot Water	Topping	152	CY
12	Accelerators	Slabs	802	CY

Formula in cell C10: = **SUM(C2:C7)**, if plasticizers specified for all concrete

Formula in cell C11: = **C7**, if only topping is done during the winter

Formula in cell C12: = **C4+C5+C7**, if accelerators specified for slabs

Analyzing Suppliers' Quotes

	A	B	C	D	E	F	G
1	Concrete Items	QTY	Unit	Supplier #1	Subtotal	Supplier #2	Subtotal
2	3000 PSI Concrete	361	CY	$115.00	$41,515	$120.00	
3	3500 PSI Concrete	417	CY	$125.00	$52,125	$130.00	
4	4000 PSI Concrete	470	CY	$130.00	$61,100	$128.00	
5	Plasticizers	1,248	CY	$9.00	$11,233	$9.00	
6	Hot Water	152	CY	$8.00	$1,218	$10.00	
7	Accelerators	802	CY	$6.00	$4,814	$8.00	
8							
9		Subtotal			$172,005		
10							
11		Sales Tax	6%		$10,320		
12							
13		Total Concrete Material			$182,325		

1. Quantities in column B are summary results from previous calculations. For example, 3,000 psi concrete was the total of SOG and slab topping, while 3,500 psi concrete was adding up the volumes of footings and walls.

2. Unit prices in columns D and F are from quotes of two suppliers. Prices include environmental fees, additional charges for small aggregates (if applicable).

3. For supplier #1, formulas in column E are as follows:

    In cell E2: = **B2*D2** (i.e., Quantity × Unit Price from Supplier #1)

    In cells E3 to E7 formulas are similar. (You can copy the formula from cell E2.)

    In cell E9: = **SUM(E2:E8),** total pretax price for Supplier #1

    In cell E11: = **E9*C11,** the material tax for Supplier #1

    In cell E13: = **E9+E11,** the final price with tax for Supplier #1

Now for Supplier #2, can you simply copy formulas from column E to column G? Unfortunately, no. If you try to do this, you will get the result of $13,800 for cell G2, which is taking cell D2 and cell F2 (the unit prices from two suppliers) and multiplying them together. Apparently that does not make sense.

The error occurs because Excel does not understand estimating. It only understands worksheet cells. In cell E2, when you entered = **B2*D2**, Excel did not think that as "Quantity × Unit Price." Instead it saw that as "Go 3 columns to the left of *the current cell*; take its value (i.e., cell B2). Next go 1 column to the left of *the current cell*, take its value (i.e. cell D2). Now multiply the two." So everything is interpreted in relation to the current cell position (i.e., cell E2). In Excel, this is called as "relative reference." When cell G2 is current cell, the old cell "B2" in the same formula is now actually referring to D2, and the old cell "D2" in the formula becomes F2. So the new formula becomes = D2*F2, which is incorrect.

There are two ways to fix this error.

1. Change the formula to "absolute reference," which will keep column B fixed. It works like this: type in cell E2, = **$B2*D2**. When you copy it over to cell G2, it becomes = **$B2*F2**. The dollar sign was to make sure column B (the quantity column) will not move, regardless of where you copy the formula.

2. Manually enter correct pricing formulas in column G for Supplier #2. This is a simple solution recommended for Excel beginners.

	A	B	C	D	E	F	G
1	**Concrete Items**	**QTY**	**Unit**	**Supplier #1**	**Subtotal**	**Supplier #2**	**Subtotal**
2	3000 PSI Concrete	**361**	CY	$115.00	**$41,515**	$120.00	**$43,320**
3	3500 PSI Concrete	**417**	CY	$125.00	**$52,125**	$130.00	**$54,210**
4	4000 PSI Concrete	**470**	CY	$130.00	**$61,100**	$128.00	**$60,160**
5	Plasticizers	**1,248**	CY	$9.00	**$11,233**	$9.00	**$11,233**
6	Hot Water	**152**	CY	$8.00	**$1,218**	$10.00	**$1,523**
7	Accelerators	**802**	CY	$6.00	**$4,814**	$8.00	**$6,418**
8							
9			Subtotal		$172,005		$176,864
10							
11			Sales Tax	6%	$10,320		$10,612
12							
13			**Total Concrete Material**		**$182,325**		**$187,476**

If entered manually one by one, correct formulas in column G are as follows:

In cell G2: = **B2*F2** (i.e., Quantity × Unit Price of Supplier #2).

In cells G3 to G7, formulas are similar. (You can copy the formula from cell G2.)

In cell G9: = **SUM(G2:G8),** total pretax price for Supplier #2.

In cell G11: = **G9*C11,** the material tax for Supplier #2.

In cell G13: = **G9+G11,** the final price with tax for Supplier #2.

Conclusion

It seems overall that Supplier #1 is cheaper than Supplier #2 by approximately $5,000, but Supplier #1 is more expensive in 4,000 psi concrete. You can try to negotiate a contract with supplier #1.

Pricing Lumber Material

Waste to Lumber and Plywood

	A	B	C	D	E	F	G
1	Waste Factor						
2	Lumber	20%					
3	Plywood	15%					
4							
5	Dimension	QTY	Unit	Waste	Unit	Adjusted QTY	Unit
6	1 x 4	1,461	BF	292	BF	1,753	BF
7	2 x 2	2,953	BF	591	BF	3,544	BF
8	2 x 4	30,754	BF	6,151	BF	36,905	BF
9	2 x 4 Studs 9'	21,859	BF	4,372	BF	26,231	BF
10	PT 2 x 4	668	BF	134	BF	802	BF
11	2 x 6	55,184	BF	11,037	BF	66,221	BF
12	2 x 6 Studs 9'	19,179	BF	3,836	BF	23,015	BF
13	PT 2 x 6	4,445	BF	889	BF	5,334	BF
14	2 x 8	4,781	BF	956	BF	5,737	BF
15	2 x 10	9,343	BF	1,869	BF	11,212	BF
16	2 x 12	5,513	BF	1,103	BF	6,616	BF
17	6 x 6	1,022	BF	204	BF	1,226	BF
18							
19	1/2" Wall Sheathing	28,893	SF	4,334	SF	33,227	SF
20	5/8" Flr Sheathing	28,132	SF	4,220	SF	32,352	SF
21	5/8" Roof Sheathing	8,229	SF	1,234	SF	9,463	SF

1. Give cell B2 a name "LumberWaste," and cell B3, "PlywoodWaste." (Refer to the Exercise four on how to name cells.)

2. Numbers in column B are your net quantities without waste.

3. Enter formula = **B6*LumberWaste** in cell D6. This is waste for lumber. (If you do not know how to name cells, enter = **B6*$B$2**. Note that you need to include the two dollar signs in the formula.)

4. Copy the formula in cell D6 all the way down to cell D17.

5. Enter formula = **B19*PlywoodWaste** in cell D19 (i.e., waste for plywood). Again, if you do not know how to name cells, enter = **B19*$B$3**.

6. Copy the formula in cell D19 to cells D20 and D21.

7. Enter formula = **B6+D6** in cell F6 (i.e., Adjusted QTY = Net QTY + Waste).

8. Copy formula in cell F6 to the rest of cells in column F.

Quotes on Dimensional Lumber

If dimensional lumbers are quoted in board feet or 1,000 board feet (mbf), then do not worry about conversions. But if they are quoted in pieces, some calculations are necessary to reach unit prices by board feet.

	A	B	C	D	E	F
1	Items	Quote	Piece LF	Factor (BF/LF)	Piece BF	Mat'l per BF
2	1 x 4	$1.24	8	0.33	2.64	$0.47
3	2 x 4	$2.65	14	0.67	9.38	$0.28
4	2 x 6	$3.90	14	1.00	14.00	$0.28
5	2 x 8	$7.15	16	1.33	21.28	$0.34
6	2 x 10	$9.76	16	1.67	26.72	$0.37
7	PT 2 x 4	$7.49	14	0.67	9.38	$0.80
8	PT 2 x 6	$11.16	14	1.00	14.00	$0.80
9	PT 4 x 4	$17.87	12	1.33	15.96	$1.12
10	PT 6 x 6	$36.75	10	3.00	30.00	$1.23

*   Prices per piece are listed in column B for lumbers with different dimensions and grades.

*   Lengths of each piece (linear foot) are listed in column C.

*   For the given dimension, column D shows board foot conversion factors.

*   Formula in cell E2: = **C2*D2** calculates how many board feet in that piece of lumber based on its dimension and length.

*   Formula in cell F2: = **B2/E2** calculates the unit price per board foot.

*   Formulas in cells E2 and F2 should be copied to the rest of the columns.

Quotes on Plywood

Plywood is normally quoted per sheet and the typical sheet size is 4 ft × 8 ft. So the math is simple: Take the sheet price and divide it by 32. In case the size is other than 4 × 8 (4 × 9 or 4 × 10), you can set up an Excel worksheet.

	A	B	C	D	E	F
1	Items	Quote	Length	Width	Area	Mat'l per SF
2	Plywood 1/2"	$15.00	8	4	32	$0.47
3	Plywood 3/4"	$26.00	8	4	32	$0.81
4	Plywood 5/8"	$22.00	8	4	32	$0.69

Formula in cell E2: = **C2*D2** (i.e., Sheet Area = Length × Width)

Formula in cell F2: = **B2/E2** (i.e., Price per Square Foot = Sheet Price/Sheet Area)

Formula in cells E2 and F2 should be copied down to the next rows.

Like dimensional lumber, it is important to note different grades of plywood (e.g., CDX, AC, FRT, PT, OSB, and T&G) as well as their thickness (e.g., ½ inch, 5/8 inch, ¾ inch). Often cost-saving ideas, such as using a different type of plywood sheets, could be proposed to the owner to help secure the deal.

Evaluating Lumber Material Quotes

	A	B	C	D	E	F	G
1	Dimension	QTY	Unit	Supplier #1	Subtotal	Supplier #2	Subtotal
2	1 x 4	1,753	BF	$0.56	$982	$0.50	
3	2 x 2	3,544	BF	$0.50	$1,772	$0.71	
4	2 x 4	36,905	BF	$0.38	$14,024	$0.36	
5	2 x 4 Studs 9'	26,231	BF	$0.38	$9,968	$0.37	
6	PT 2 x 4	802	BF	$0.85	$681	$0.70	
7	2 x 6	66,221	BF	$0.34	$22,515	$0.34	
8	2 x 6 Studs 9'	23,015	BF	$0.42	$9,666	$0.41	
9	PT 2 x 6	5,334	BF	$0.77	$4,107	$0.72	
10	2 x 8	5,737	BF	$0.35	$2,008	$0.35	
11	2 x 10	11,212	BF	$0.37	$4,148	$0.39	
12	2 x 12	6,616	BF	$0.34	$2,249	$0.33	
13	6 x 6	1,226	BF	$1.00	$1,226	$0.90	
14							
15	1/2" Wall Sheathing	33,227	SF	$0.48	$16,038	$0.50	
16	5/8" Flr Sheathing	32,352	SF	$0.74	$23,791	$0.70	
17	5/8" Roof Sheathing	9,463	SF	$0.74	$6,959	$0.65	
18							
19				Subtotal	$120,135		
20							
21			Sales Tax	6%	$7,208	6%	
22							
23			Total framing material	$127,343			

Numbers in column B are adjusted quantities including wastes.

Numbers in columns D and F are unit prices calculated from material quotes.

For Supplier #1 in column E:

1. Enter formula in cell E2: = **$B2*D2** (i.e., Quantity × Unit Price).
2. Copy the formula in cell E2 all the way down to cell E17.
3. Enter formula in cell E19: = **SUM(E2:E17)**, total pretax price.
4. Enter formula in cell E21: = **E19*$D21**, material tax.
5. Enter formula in cell E23: = **SUM(E19:E21)**, total price for Supplier #1.

Now, can formulas in column E be copied to column G for Supplier #2? Yes, it can!

	A	B	C	D	E	F	G
1	Dimension	QTY	Unit	Supplier #1	Subtotal	Supplier #2	Subtotal
2	1 x 4	1,753	BF	$0.56	$982	$0.50	$877
3	2 x 2	3,544	BF	$0.50	$1,772	$0.71	$2,502
4	2 x 4	36,905	BF	$0.38	$14,024	$0.36	$13,286
5	2 x 4 Studs 9'	26,231	BF	$0.38	$9,968	$0.37	$9,705
6	PT 2 x 4	802	BF	$0.85	$681	$0.70	$562
7	2 x 6	66,221	BF	$0.34	$22,515	$0.34	$22,320
8	2 x 6 Studs 9'	23,015	BF	$0.42	$9,666	$0.41	$9,333
9	PT 2 x 6	5,334	BF	$0.77	$4,107	$0.72	$3,852
10	2 x 8	5,737	BF	$0.35	$2,008	$0.35	$2,026
11	2 x 10	11,212	BF	$0.37	$4,148	$0.39	$4,379
12	2 x 12	6,616	BF	$0.34	$2,249	$0.33	$2,194
13	6 x 6	1,226	BF	$1.00	$1,226	$0.90	$1,109
14							
15	1/2" Wall Sheathing	33,227	SF	$0.48	$16,038	$0.50	$16,561
16	5/8" Flr Sheathing	32,352	SF	$0.74	$23,791	$0.70	$22,646
17	5/8" Roof Sheathing	9,463	SF	$0.74	$6,959	$0.65	$6,142
18							
19				Subtotal	$120,135		$117,494
20							
21			Sales Tax	6%	$7,208	6%	$7,050
22							
23			Total framing material		$127,343		$124,544

What made it work are the dollar signs for formulas in column E. For example, in cell E2: = **$B2*D2**. This formula only takes the values from column B (i.e., quantities), regardless of where the current cell is located. In Excel, this is called as "absolute reference."

Conclusion

It appears that Supplier #2 is slightly cheaper.

Evaluating Joist/Truss Suppliers

Suppliers for preengineered wood joists and trusses are evaluated in a similar way to subcontractors, except they do not install. Base bid prices are to be adjusted with a list of factors to ensure they furnish everything needed for framing material.

	A	B	C	D	E
1	Name	Plug Price	Supplier 1	Supplier 2	Supplier 3
2					
3	Joist/Beam Supply		$60,000	$72,000	$67,000
4	TJI		Included	Included	Included
5	Parallam		Included	Included	Included
6	Timberstrand		Included	Included	Included
7	Glulam	Add $10,000	$10,000	$10,000	Included
8	Joist Hanger		Included	Included	Included
9	Connector	Add $5,000	$5,000	Included	$5,000
10	Joist/Beam Subtotal		$75,000	$82,000	$72,000
11					
12	Roof Truss Supply		$57,000	$58,000	$62,000
13					
14	Total incl. Truss		$132,000	$140,000	$134,000

1. Base bids for joist/beam supply are in row 3, and truss supply in row 12.
2. Adjusting factors for joist/beam are listed in row 4 to row 9.
3. Formula in cell C10: = **SUM(C3:C9)**, base bid plus all adds for Supplier #1.
4. Copy formula in cell C10 to cells D10 and E10.
5. Formula in cell C14: = **C10+C12**, adding joist/beam and truss for Supplier #1.
6. Copy formula in cell C14 to cells D14 and E14.

Conclusion

It appears Supplier #1 is low overall, but others are fairly close. Supplier #3 actually beat Supplier #1 on wood joists and beams.

Pricing Concrete Labor

Calculating Concrete Crew Rate

1. Give cell B1 the name "Burden" for the rate you decided. (Refer to Exercise Four on how to name cells.)
2. List crew members in column A and their basic wages in column B.
3. Formula in cell C4: = **B4*Burden** (i.e., Basic Wage × Burden Rate). If you do not know how to name cells, enter = **B4*$B$1**. Note the two dollar signs.

	A	B	C	D
1	Burden Rate	30%		
2				
3	Crew Members	Basic Wage	Burden	Adjusted Rate
4	Foreman	$35.00	$10.50	$45.50
5	Journeyman	$30.00	$9.00	$39.00
6	Laborer	$25.00	$7.50	$32.50
7	Crew Rate			$39.00

4. Formula in cell D4: = **B4+C4** (i.e., Basic Wage + Labor Burden).
5. Copy formula in cells C4 and D4 down to the next two rows.
6. Formula in cell D7: = **AVERAGE(D4:D6)** for overall crew rate.

Note that AVERAGE worksheet function is used to calculate the average value for a list of numbers. To use it, you can enter a cell range such as (D4:D6), or simply enter cells one after another such as (D4, D5, D6)

Estimating Concrete Labor Costs

	A	B	C	D	E	F	G
1	Items	QTY	UNIT	Labor			
2				Man-hour per Unit	Man-hours	Crew Rate	Subtotal
3	Form strip footing	175	SF	0.10	17.5	$ 39.00	$ 683
4	Form pads	48	SF	0.10	4.8	$ 39.00	$ 187
5	Form foundation wall	520	SF	0.15	78	$ 39.00	$ 3,042
6							
7	Placing strip footing	3.7	CY	1.00	3.7	$ 39.00	$ 144
8	Placing pads	1.1	CY	1.20	1.3	$ 39.00	$ 51
9	Placing foundation wall	7.5	CY	1.00	7.5	$ 39.00	$ 293
10							
11	Placing reinforcing	340	Lbs	0.03	10.2	$ 39.00	$ 398
12							
13	Total Labor Cost						$ 4,798

Formula in cell E3: = **B3*D3** (i.e., Quantity × Man-hour per Unit)

Formula in cell G3: = **E3*F3** (i.e., Man-hours × Crew Rate)

Pricing Framing Labor

Calculating Framing Crew Rate

	A	B	C	D
1	Burden Rate	25%		
2				
3	Crew Members	Basic Wage	Burden	Adjusted Rate
4	Foreman	$35.00	$8.75	$43.75
5	Carpenters	$30.00	$7.50	$37.50
6	Laborer 1	$28.00	$7.00	$35.00
7	Laborer 2	$25.00	$6.25	$31.25
8	Average crew rate			$36.88

Estimating Framing Labor Costs

Formulas in these two worksheets are similar to the ones under "Pricing Concrete Labor." See previous example for details.

	A	B	C	D	E	F	G
1				Labor			
2	Items	QTY	UNIT	Man-hour per Unit	Man-hours	Crew Rate	Subtotal
3	Framing lumber	5,500	BF	0.03	165	$36.88	**$6,085**
4	Framing plywood	2,500	SF	0.01	25	$36.88	**$922**
5	Install roof trusses	21	EA	1.50	31.5	$36.88	**$1,162**
6	Install TJI/beams	42	EA	1.00	42	$36.88	**$1,549**
7	Install wood stairs	1	EA	3.00	3	$36.88	**$111**
8	Framing openings	5	EA	2.00	10	$36.88	**$369**
9	Backframing	1,000	SF	0.05	50	$36.88	**$1,844**
10							
11	Total Labor Costs						$12,041

Combined Labor/Material Pricing

Sometimes for straight-forward items, it may be more convenient to set up a combined labor/material pricing worksheet, as follows:

	A	B	C	D	E	F	G	H	I	J
1	Item	QTY	Unit	Material		Labor				Total Costs
2				Unit Price	Subtotal	Man-hour per Unit	Man-hrs	Crew Rate	Subtotal	
3	Wood Door	800	EA	$350.00	$280,000	3.20	2,560	$31.00	$79,360	$359,360
4										
5										

Formulas are as follows:

In cell E3: = **B3*D3** (i.e., Material Cost = Quantity × Material Unit Price)

In cell G3: = **B3*F3** (i.e., Man-hours = Quantity × Man-hour per Unit)

In cell I3: = **G3*H3** (i.e., Labor Cost = Man-hours × Crew Rate)

In cell J3: = **E3+I3** (i.e., Total Cost = Material Cost + Labor Cost)

For each door, the unit cost is $359,360/800 = $450 to supply and install.

Evaluating Concrete Contractors

	A	B	C	D	E
1	**Formwork**	**Plug Price**	Sub 1	Sub 2	Sub 3
2					
3	**Base Price**		$50,000	$35,000	$60,000
4					
5	Forming		INCL	INCL	INCL
6	Concrete material		NIC	NIC	NIC
7	Concrete placing		INCL	INCL	INCL
8	Concrete finishing	Add $5,000	$5,000	$5,000	$5,000
9	Rebar supply or install		NIC	NIC	NIC
10	Excavation/backfill		NIC	NIC	NIC
11	Fine grading	Add $2,000	$2,000	$2,000	INCL
12	Soil treatment		NIC	NIC	NIC
13					
14	Footings/pads		INCL	INCL	INCL
15	Fdn walls		INCL	INCL	INCL
16	Slab on grade	Add $10,000	$10,000	$10,000	INCL
17	Upper floor topping	Add $5,000	$5,000	$5,000	INCL
18	Beams & columns	Add $8,000	INCL	$8,000	INCL
19	Stairs		INCL	INCL	INCL
20	Building sidewalks	Add $3,000	$3,000	$3,000	$3,000
21	Site concrete	Add $7,000	$7,000	$7,000	$7,000
22					
23	Layout		INCL	INCL	INCL
24	Scaffolding	Add $3,000	INCL	$3,000	INCL
25	Forklift/crane	Add $4,000	$4,000	$4,000	INCL
26	Expansion joints	Add $500	$500	$500	$500
27	PVC water stop	Add $250	$250	$250	$250
28	Chamfer strip	Add $750	$750	$750	$750
29	Form openings	Add $1,200	$1,200	INCL	$1,200
30	Patching exposed wall	Add $400	$400	$400	$400
31	M/E equipment pads	Add $1,500	$1,500	$1,500	$1,500
32	Embeds	Add $600	$600	INCL	INCL
33	Rebar unloading	Add $200	INCL	INCL	$200
34					
35	**Adjusted total**		$91,200	$85,400	$79,800

"Plug prices" are inputs from subs, or your own estimate for what they excluded. Formula in cell C35: = **SUM(C3:C33)**, which is base bid plus all adjustments.

Evaluating Wood Framers

	A	B	C	D	E
1	Framers	Plug Price	Sub 1	Sub 2	Sub 3
2					
3	Base Price		$100,000	$90,000	$120,000
4					
5	Framing material		NIC	NIC	NIC
6	Floor framing and sheathing		INCL	INCL	INCL
7	Wall framing and sheathing		INCL	INCL	INCL
8	Roof framing and sheathing		INCL	INCL	INCL
9	Fascia and soffit	Add $5,000	INCL	$5,000	INCL
10	Framing layout	Add $1,000	INCL	$1,000	INCL
11	Nails/Screws	Add $5,000	INCL	$5,000	INCL
12	Power	Add $6,000	$6,000	$6,000	$6,000
13	Forklift	Add $7,000	INCL	$7,000	INCL
14	Crane for joist/truss	Add $15,000	INCL	$15,000	$15,000
15	Backframing	Add $10,000	$10,000	$10,000	$10,000
16	Wood stairs	Add $2,000	INCL	INCL	$2,000
17					
18	Adjusted Price		$116,000	$139,000	$153,000
19					
20	Lowest Bid	$116,000			

Formula in cell C18: = **SUM(C3:C16)** (i.e., base bid plus all adjustments)

Formulas in cells D18 and E18 are similar to the one in cell C18. (You can copy the formula from cell C18 over.)

Formula in cell B20: = **MIN(C18:E18)**, to pick the lowest bid

MIN worksheet function is used to decide the lowest of many numbers. Its opposite is MAX worksheet function, which picks the highest number.

EXERCISE 7: BUDGET ANALYSIS

Mission Briefing

This exercise starts with the cost summary sheet you completed in Exercise Two. Then you figure markup (i.e., office overhead and profit) based on the sum of all direct costs. Finally, you can do an analysis on the total budget.

Skill Level: Easy/Intermediate

Suggestions to Readers

This last exercise is relatively easy and interesting. In addition to reviewing the skills you already know, you will learn about how to hide or unhide contents, write notes (comments), link values in two different worksheets, and draw charts or graphs for data analysis. It's easier than you think!

Summarizing Costs

Find the cost summary sheet you did in Exercise 2 and open it up.

	A	B	C	D	E	F	G	H
1			Simple Life Residence Cost Summary					
2	Ref #	Cost Items	Estimated Costs	Draw 1	Draw 2	Draw 3	Actual Costs	Cost Variance
3	001	Jobsite Overhead	$7,000	$4,000	$2,500	$500	$7,000	$0
4	002	Excavation	$1,500	$800	$0	$500	$1,300	$200
5	003	Foundation	$6,000	$7,500	$0	$0	$7,500	-$1,500
6	004	Framing	$20,000	$0	$18,000	$2,500	$20,500	-$500
7	005	Roof & Siding						
8	006	Insulation						
9	007	Drywall						
10	008	Doors & Windows						
11	009	Cabinets						
12	010	Accessories						
13	011	Painting						
14	012	Floor Coverings						
15	013	Electrical						
16	014	Plumbing and Heating						
17	015	Office Overhead						
18	016	Profit						
19	017	Contirgency						

1. Select columns D through H.
2. Right-click your mouse. Pick *Hide* (picking *Unhide* will restore the previously hidden columns or rows, if any).

D	E	F	G	H	I
Simple Life Residence Cost Summary					
Draw 1	Draw 2	Draw 3	Actual Costs	Cost Variance	
$4,00C	$2,500	$500	$7,000	$0	
$80C	$0	$500	$1,300	$200	
$7,50C	$0	$0	$7,500		
$C	$18,000	$2,500	$20,500		

Context menu options:
- Cut
- Copy
- Paste
- Paste Special...
- Insert
- Delete
- Clear Ccntents
- Format Cells...
- Column Width...
- **Hide**
- Unhide

	A	B	C
1		Simple Life Residence Cost	
2	Ref #	Cost Items	Estimated Costs
3	001	Jobsite Overhead	$7,000
4	002	Excavation	$1,500
5	003	Foundation	$6,000
6	004	Framing	$20,000
7	005	Roof & Siding	$15,000
8	006	Insulation	$3,000
9	007	Drywall	$7,000
10	008	Doors & Windows	$3,000
11	009	Cabinets	$5,000
12	010	Accessories	$1,000
13	011	Painting	$4,000
14	012	Floor Coverings	$4,500
15	013	Electrical	$8,000
16	014	Plumbing and Heating	$13,000

3. Now you have a simplified budget worksheet for this exercise.

4. Enter cost information for the rest of the worksheet.

5. Select row 17, right-click your mouse and pick *Insert* to insert a new row.

6. In new cell A17, type **Total Cost.**

7. Select both cells A17 and B17, click Merge button [icon] on the toolbar.

8. In new cell C17, click the AutoSum button Σ on the toolbar to add all costs.

	A	B	C
15	013	Electrical	$8,000
16	014	Plumbing and Heating	$13,000
17	Total Cost		$98,000
18	015	Office Overhead	
19	016	Profit	

Note that for sake of simplicity, this exercise does not consider any soft costs.

Estimating Markup

Continue using the same worksheet.

1. Enter formula in cell C18: **=C17*4%.**

	A	B	C
17	**Total Cost**		**$98,000**
18	015	Office Overhead	$3,920
19	016	Profit	
20	017	Contingency	

2. With cell C18 selected, press SHIFT + F2 keys to add some explanations (i.e., "comments") for what the formula actually means.

3. In the small box that appears, type **Overhead rate 4%** to remind yourself.

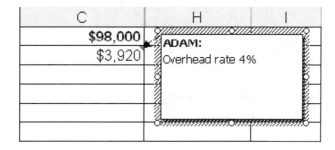

4. With cell C18 still selected, right-click and pick *Hide Comment*.

5. In cell C19, enter = **C17*10%.** Put similar comments for profit rate (e.g., put in "profit is 10%").

6. In cell C20, enter = **C17*5%.** Put similar comments to explain contingencies.

7. In cell A21, type **Total Budget**.

8. In cell C21, enter = **SUM(C17:C20)** to add up everything.

	A	B	C
17	**Total Cost**		**$98,000**
18	C15	Office Overhead	$3,920
19	C16	Profit	$9,800
20	C17	Contingency	$4,900
21	**Total Budget**		**$116,620**

9. The comments for cells C18 to C20 are hidden by default. If you move your mouse over any of these cells, a comment will display. If you move the mouse away, it will disappear.

	A	B	C	H	I
17	Total Cost		$98,000		
18	015	Office Overhead	$3,920		
19	016	Profit	$9,800	ADAM:	
20	017	Contingency	$4,900	5% for unclear design	
21	Total Budget		$116,620		
22					
23					
24					

Cost Analysis

Continue using the same workbook file. You will now work on two worksheets.

1. Press SHIFT + F11 keys to insert a new worksheet.
2. Rename it as "Analysis." Enter text fields as shown.

	A	B
1	Budget Analysis	
2	Cost	
3	Office Overhead	
4	Profit	
5	Contingency	
6	Total	

3. To get values for cells in column B, you need to link this new sheet with the original "Cost Summary" worksheet.

4. Go back to "Cost Summary" sheet, and find the cell containing the total cost (i.e., cell C17). Press CTRL + C to copy its value.

17	Total Cost		$98,000
18	015	Office Overhead	$3,920
19	016	Profit	$9,800
20	017	Contingency	$4,900
21	Total Budget		$116,620
22			

◄ ◄ ► ►◄ \ **Cost Summary** ⟨ Analysis /

5. Go back to "Analysis" sheet, select cell B2. Right-click, pick *Paste Special*.

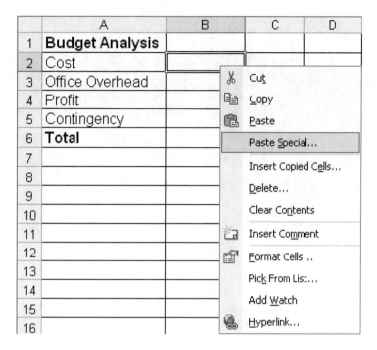

6. In the dialog that appears, simply click *Paste Link* button (do NOT click *OK*).

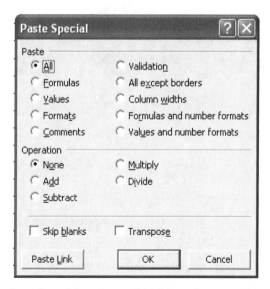

7. Now cell B2 is linked to the value of "total costs" in "Cost Summary" worksheet. Every time the budget changes, the revised total cost will be automatically transferred to this sheet.

8. Do the same thing for cells B3, B4, and B5. That is, copy cells from "Cost Summary" sheet and paste links to this new sheet.

	A	B
1	**Budget Analysis**	Value
2	Cost	$ 98,000
3	Office Overhead	$ 3,920
4	Profit	$ 9,800
5	Contingency	$ 4,900
6	**Total**	**$116,620**

9. Enter = **SUM(B2:B5)** in cell B6 to add everything. It should be the same budget number in "Cost Summary" sheet.

10. Select cells A2 to B5 (i.e., names and their values). Click the chart wizard button on the toolbar.

11. In step 1, choose *Pie* for *Chart Type*. Click *OK*.

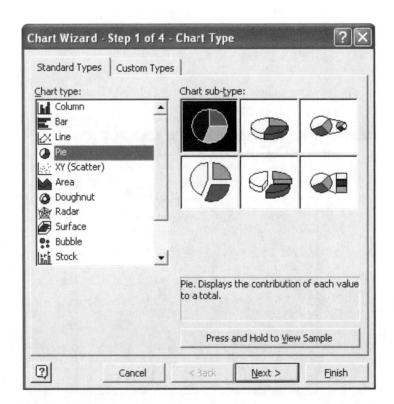

12. Step 2 shows what data the chart is based on. Click *Next*.

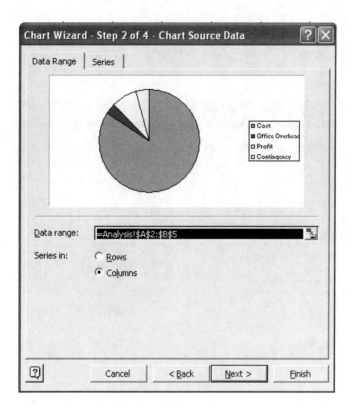

13. In step 3, give your chart a name (e.g., **Analysis**). Do NOT click *Next* yet.

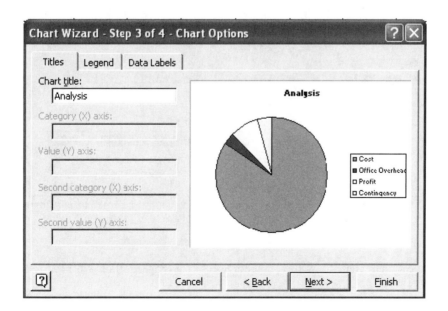

14. Still in Step 3, click one of the three tabs *Data Labels* on the top.

15. Under *Label Contains,* check *Percentage*. Now click *Next*.

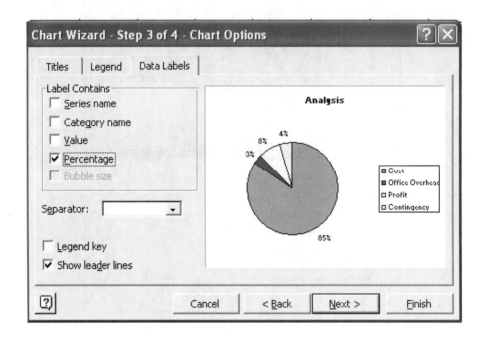

16. In the final step, you can decide where to put the chart. Then click *Finish*.

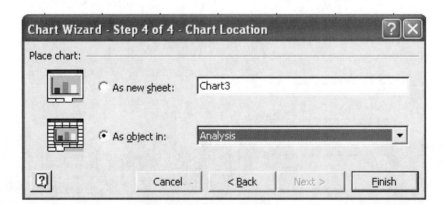

17. Now you have a chart that shows the percentages of cost (85%), office overhead (3%), profit (8%), and contingency (4%) in relation to *the total budget*.

	A	B	C	D	E	F	G	H
1	**Budget Analysis**	Value						
2	Cost	$ 98,000						
3	Office Overhead	$ 3,920						
4	Profit	$ 9,800						
5	Contingency	$ 4,900						
6	**Total**	**$116,620**						
7								

Analysis

Legend: ■ Cost ■ Office Overhead ⊓ Profit □ Contingency

85% 8% 3% 4%

Note that the percentages you entered earlier for overhead (4%), profit (10%), and contingency (5%) were in relation to *costs only*. For example, overhead was calculated as 4% of the direct cost: 4% × $98,000 = $3,920. But in the chart, overhead is $3,920/$116,620 = 3% of the total budget.

When your budget values change, the chart will be automatically updated with new percentages. If you like, you can even copy the chart and paste it into a Word document.

APPENDIX

COMMON UNIT CONVERSIONS

Convert From	Multiply By	To Obtain
Acres	0.4047	Hectares
Acres	43,560	Square Feet
Acres	0.0016	Square Miles
Acres	4,047	Square Meters
Acres	0.0041	Square Kilometers
Acres	4,840	Square Yards
Acre-feet	43,560	Cubic Feet
Acre-feet	1,233	Cubic Meters
Acre-feet	1,613	Cubic Yards
Acre-feet	325,900	Gallons (US)
Acre-inches	3,630	Cubic Feet
Acre-inches	102.79	Cubic Meters
Acre-inches	134.44	Cubic Yards
Acre-inches	27,154	Gallons (US)
Atmospheres	76	Centimeters of Mercury
Atmospheres	29.92	Inches of Mercury
Atmospheres	1,033	Centimeters of Water
Atmospheres	33.9	Feet of Water

Convert From	Multiply By	To Obtain
Atmospheres	101.325	Kilopascals
Atmospheres	101,325	Pascals
Atmospheres	14.7	Pounds per Square Inch
British Thermal Units	252.16	Calories
British Thermal Units	778.17	Foot-Pounds
British Thermal Units	0.0004	Horsepower-hours
British Thermal Units	1,055	Joules
British Thermal Units	0.252	Kilogram-calories
British Thermal Units	107.51	Kilogram-meters
British Thermal Units	0.0003	Kilowatt-Hours
Calories	0.00397	British Thermal Units
Calories	4.184	Joules
Calories	3.086	Foot-Pounds
Centimeters	10	Millimeters
Centimeters	0.01	Meters
Centimeters	0.394	Inches
Centimeters	0.033	Feet
Centimeters	0.011	Yards
Cubic Feet	28.32	Liters
Cubic Feet	0.0283	Cubic Meters
Cubic Feet	1,728	Cubic Inches
Cubic Feet	957.51	Fluid Ounces (US)
Cubic Feet	59.844	Pints (US)
Cubic Feet	29.922	Quarts (US)
Cubic Feet	7.481	Gallons (US)
Cubic Feet	0.037	Cubic Yards
Cubic Feet per Second	448.83	Gallons (US) per Minute
Cubic Feet per Second	26,930	Gallons (US) per Hour
Cubic Feet per Minute	0.125	Gallons (US) per Second
Cubic Feet per Minute	448.83	Gallons (US) per Hour

Convert From	Multiply By	To Obtain
Cubic Feet per Hour	0.002	Gallons (US) per Second
Cubic Feet per Hour	0.125	Gallons (US) per Minute
Cubic Inch	0.016	Liters
Cubic Inch	1.64×10^{-5}	Cubic Meters
Cubic Inch	0.554	Fluid Ounces (US)
Cubic Inch	0.035	Pints (US)
Cubic Inch	0.017	Quarts (US)
Cubic Inch	0.004	Gallons (US)
Cubic Inch	0.001	Cubic Feet
Cubic Inch	2.14×10^{-5}	Cubic Yards
Cubic Meters	1,000	Liters
Cubic Meters	264.172	Gallons (US)
Cubic Meters	35.315	Cubic Feet
Cubic Meters	1.308	Cubic Yards
Cubic Yards	764.555	Liters
Cubic Yards	0.765	Cubic Meters
Cubic Yards	46,656	Cubic Inches
Cubic Yards	201.974	Gallons (US)
Cubic Yards	27	Cubic Feet
Degrees (angle)	0.01745	Radians
Degrees (angle)	0.00278	Circles
Degrees (angle)	60	Minutes
Feet	304.8	Millimeters
Feet	30.48	Centimeters
Feet	0.305	Meters
Feet	3.05×10^{-4}	Kilometers
Feet	12	Inches
Feet	0.333	Yards
Feet	1.89×10^{-4}	Miles (statute)
Feet	1.65×10^{-4}	Miles (nautical)

Convert From	Multiply By	To Obtain
Feet of Air	0.0009	Feet of Mercury
Feet of Air	0.00122	Feet of Water
Feet of Air	0.00108	Inches of Mercury
Feet of Air	0.00053	Pounds per Square Inch
Feet of Mercury	30.48	Centimeters of Mercury
Feet of Mercury	13.6086	Feet of Water
Feet of Mercury	163.3	Inches of Water
Feet of Mercury	5.8938	Pounds per Square Inch
Feet of Water	0.0295	Atmospheres
Feet of Water	2.2419	Centimeters of Mercury
Feet of Water	0.8826	Inches of Mercury
Feet of Water	304.78	Kilograms per Square Meter
Feet of Water	2,988.9	Pascals
Feet of Water	62.424	Pounds per Square Foot
Feet of Water	0.4335	Pounds per Square Inch
Feet per Second	0.305	Meters per Second
Feet per Second	1.097	Kilometers per Hour
Feet per Second	0.592	Knots
Feet per Second	0.682	Miles (statute) per Hour
Foot-Pounds	0.00129	British Thermal Units
Foot-Pounds	1.356	Joules
Foot-Pounds	0.324	Calories
Foot-Pounds	3.766×10^{-7}	Kilowatt-Hours
Gallons (US)	3.785	Liters
Gallons (US)	0.00379	Cubic Meters
Gallons (US)	231	Cubic Inches
Gallons (US)	128	Fluid Ounces (US)
Gallons (US)	8	Pints (US)
Gallons (US)	4	Quarts (US)
Gallons (US)	0.134	Cubic Feet

Convert From	Multiply By	To Obtain
Gallons (US)	4.95×10^{-3}	Cubic Yards
Gallons (US) per Second	8.021	Cubic Feet per Minute
Gallons (US) per Second	481.25	Cubic Feet per Hour
Gallons (US) per Minute	2.23×10^{-3}	Cubic Feet per Second
Gallons (US) per Minute	8.021	Cubic Feet per Hour
Grams	1,000	Milligrams
Grams	0.001	Kilograms
Grams	0.0353	Ounces
Grams	2.20×10^{-3}	Pounds
Hectares	10,000	Square Meters
Hectares	0.01	Square Kilometers
Hectares	107,639	Square Feet
Hectares	11,960	Square Yards
Hectares	2.471	Acres
Hectares	3.86×10^{-3}	Square Miles
Horsepower (US)	42.375	BTU per Minute
Horsepower (US)	2,543	BTU per Hour
Horsepower (US)	550	Foot-Pounds per Second
Horsepower (US)	33,000	Foot-Pounds per Minute
Horsepower (US)	1.014	Horsepower (Metric)
Horsepower (US)	0.7457	Kilowatts
Horsepower (US)	745.7	Watts
Inches	25.4	Millimeters
Inches	2.54	Centimeters
Inches	0.0254	Meters
Inches	0.0833	Feet
Inches	0.0278	Yards
Inches of Mercury	0.0334	Atmospheres
Inches of Mercury	1.133	Feet of Water
Inches of Mercury	3,386	Pascals

Convert From	Multiply By	To Obtain
Inches of Mercury	70.526	Pounds per Square Foot
Inches of Mercury	0.4912	Pounds per Square Inch
Inches of Water	2.46×10^{-3}	Atmospheres
Inches of Water	0.07355	Inches of Mercury
Inches of Water	25.398	Kilograms per Square Meter
Inches of Water	5.202	Pounds per Square Foot
Inches of Water	0.036	Pounds per Square Inch
Joules	9.5×10^{-4}	British Thermal Units
Joules	0.239	Calories
Joules	2.778×10^{-7}	Kilowatt-Hours
Joules	0.738	Foot-Pounds
Kilograms	1,000	Grams
Kilograms	35.274	Ounces
Kilograms	2.205	Pounds
Kilograms	1.1×10^{-3}	Short Tons
Kilograms	9.8×10^{-4}	Long Tons
Kilogram-calories	3.968	British Thermal Units
Kilograms per Square Meter	3.28×10^{-3}	Feet of Water
Kilograms per Square Meter	0.2048	Pounds per Square Foot
Kilograms per Square Meter	1.42×10^{-3}	Pounds per Square Inch
Kilometers	1,000	Meters
Kilometers	3,281	Feet
Kilometers	1,094	Yards
Kilometers	0.621	Miles (statute)
Kilometers per Hour	0.278	Meters per Second
Kilometers per Hour	0.54	Knots
Kilometers per Hour	0.911	Feet per Second
Kilometers per Hour	0.621	Miles (statute) per Hour
Kilopascals	9.87×10^{-3}	Atmospheres
Kilopascals	0.2952	Inches of Mercury

Convert From	Multiply By	To Obtain
Kilopascals	4.021	Inches of Water
Kilopascals	1,000	Pascals
Kilopascals	20.885	Pounds per Square Foot
Kilopascals	0.145	Pounds per Square Inch
Kilowatts	56.8725	BTU per Minute
Kilowatts	1.341	Horsepower
Kilowatt-Hours	3,412	British Thermal Units
Kilowatt-Hours	3,600,000	Joules
Kilowatt-Hours	860,421	Calories
Kilowatt-Hours	2,655,000	Foot-Pounds
Knots	0.514	Meters per Second
Knots	1.852	Kilometers per Hour
Knots	1.688	Feet per Second
Knots	1.151	Miles (statute) per Hour
Liters	0.001	Cubic Meters
Liters	61.024	Cubic Inches
Liters	33.814	Fluid Ounces (US)
Liters	0.264	Gallons (US)
Liters	0.0353	Cubic Feet
Liters	1.31×10^{-3}	Cubic Yards
Megapascals	145	Pounds per Square Inch
Meters	1,000	Millimeters
Meters	100	Centimeters
Meters	0.001	Kilometers
Meters	39.37	Inches
Meters	3.281	Feet
Meters	1.094	Yards
Meters	6.21×10^{-4}	Miles (statute)
Meters per Second	3.6	Kilometers per Hour
Meters per Second	1.944	Knots

Convert From	Multiply By	To Obtain
Meters per Second	3.281	Feet per Second
Meters per Second	2.237	Miles (statute) per Hour
Miles (nautical)	1.1516	Miles (statute)
Miles (statute)	1,609	Meters
Miles (statute)	1.609	Kilometers
Miles (statute)	5,280	Feet
Miles (statute)	1,760	Yards
Miles (statute)	0.8684	Miles (nautical)
Miles (statute) per Hour	0.447	Meters per Second
Miles (statute) per Hour	1.609	Kilometers per Hour
Miles (statute) per Hour	0.869	Knots
Miles (statute) per Hour	1.467	Feet per Second
Millimeters	0.1	Centimeters
Millimeters	0.001	Meters
Millimeters	0.0394	Inches
Millimeters	3.28×10^{-3}	Feet
Ounces	28,350	Milligrams
Ounces	28.35	Grams
Ounces	0.0283	Kilograms
Ounces	0.0625	Pounds
Ounces (Fluid, US)	0.0296	Liters
Ounces (Fluid, US)	1.805	Cubic Inches
Ounces (Fluid, US)	7.81×10^{-3}	Gallons (US)
Ounces (Fluid, US)	1.04×10^{-3}	Cubic Feet
Pascals	9.87×10^{-6}	Atmospheres
Pascals	3.35×10^{-4}	Feet of Water
Pascals	2.95×10^{-4}	Inches of Mercury
Pascals	4.01×10^{-3}	Atmospheres
Pascals	0.102	Kilograms per Square Meter
Pascals	0.021	Pounds per Square Foot

Convert From	Multiply By	To Obtain
Pascals	1.45×10^{-4}	Pounds per Square Inch
Pounds	453.592	Grams
Pounds	0.454	Kilograms
Pounds	4.54×10^{-4}	Tons (Metric)
Pounds	16	Ounces
Pounds	0.0005	Tons (Short)
Pounds	4.46×10^{-4}	Tons (Long)
Pounds per Cubic Foot	16.02	Kilograms per Cubic Meter
Pounds per Cubic Foot	5.79×10^{-4}	Pounds per Cubic Inch
Pounds per Cubic Inch	27.68	Grams per Cubic Centimeter
Pounds per Cubic Inch	27,680	Kilograms per Cubic Meter
Pounds per Cubic Inch	1,728	Pounds per Cubic Foot
Pounds per Square Foot	0.016	Feet of Water
Pounds per Square Foot	4.89	Kilograms per Square Meter
Pounds per Square Foot	0.007	Pounds per Square Inch
Pounds per Square Foot	47.88	Pascals
Pounds per Square Foot	0.048	Kilopascals
Pounds per Square Inch	2.31	Feet of Water
Pounds per Square Inch	27.73	Inches of Water
Pounds per Square Inch	0.0703	Kilograms per Square Centimeters
Pounds per Square Inch	2.036	Inches of Mercury
Pounds per Square Inch	144	Pounds per Square Foot
Pounds per Square Inch	6,895	Pascals
Pounds per Square Inch	6.895	Kilopascals
Pounds per Square Inch	0.0069	Megapascals
Radians	57.3	Degrees
Square Centimeters	100	Square Millimeters
Square Centimeters	0.0001	Square Meters
Square Centimeters	0.155	Square Inches
Square Centimeters	1.08×10^{-3}	Square Feet

Convert From	Multiply By	To Obtain
Square Centimeters	1.2×10^{-4}	Square Yards
Square Feet	0.0929	Square Meters
Square Feet	9.3×10^{-6}	Hectares
Square Feet	9.3×10^{-8}	Square Kilometers
Square Feet	144	Square Inches
Square Feet	0.111	Square Yards
Square Feet	2.3×10^{-5}	Acres
Square Inches	645.16	Square Millimeters
Square Inches	6.452	Square Centimeters
Square Inches	6.45×10^{-4}	Square Meters
Square Inches	6.94×10^{-3}	Square Feet
Square Inches	7.72×10^{-4}	Square Yards
Square Kilometers	1,000,000	Square Meters
Square Kilometers	100	Hectare
Square Kilometers	10,760,000	Square Feet
Square Kilometers	1,196,000	Square Yards
Square Kilometers	247.105	Acres
Square Kilometers	0.386	Square Miles
Square Meters	10,000	Square Centimeters
Square Meters	0.0001	Hectares
Square Meters	0.000001	Square Kilometers
Square Meters	1,550	Square Inches
Square Meters	10.764	Square Feet
Square Meters	1.196	Square Yards
Square Meters	2.47×10^{-4}	Acres
Square Miles	2,590,000	Square Meters
Square Miles	259	Hectares
Square Miles	2.59	Square Kilometers
Square Miles	27,880,000	Square Feet
Square Miles	3,098,000	Square Yards

Convert From	Multiply By	To Obtain
Square Miles	640	Acres
Square Millimeters	0.01	Square Centimeters
Square Millimeters	0.000001	Square Meters
Square Millimeters	1.55×10^{-3}	Square Inches
Square Millimeters	1.08×10^{-5}	Square Feet
Square Yards	0.836	Square Meters
Square Yards	8.36×10^{-5}	Hectares
Square Yards	8.36×10^{-7}	Square Kilometers
Square Yards	1,296	Square Inches
Square Yards	9	Square Feet
Square Yards	2.07×10^{-4}	Acres
Tons (Metric)	1,000	Kilograms
Tons (Metric)	2,205	Pounds
Tons (Metric)	1.102	Tons (Short)
Tons (Metric)	0.984	Tons (Long)
Tons (Short)	907.184	Kilograms
Tons (Short)	0.907	Tons (Metric)
Tons (Short)	2000	Pounds
Tons (Short)	0.893	Tons (Long)
Tons (Long)	1,016	Kilograms
Tons (Long)	1.016	Tons (Metric)
Tons (Long)	2,240	Pounds
Tons (Long)	1.12	Tons (Short)
Watts	3.4121	BTU per Hour
Watts	0.0568	BTU per Minute
Watts	0.0013	Horsepower
Watts	14.34	Calories per Minute
Watt-hours	3.4144	British Thermal Units
Yards	914.4	Millimeters
Yards	91.44	Centimeters

Convert From	Multiply By	To Obtain
Yards	0.914	Meters
Yards	9.14×10^{-4}	Kilometers
Yards	36	Inches
Yards	3	Feet
Yards	5.68×10^{-4}	Miles (statute)

ESTIMATING AND BIDDING GLOSSARY

addendum/addenda   A written document adding to, clarifying, or changing bidding documents. An addendum is generally issued after bid documents have been made available to contractors, but prior to bid closing. It is part of contract documents.

agreement   Specific documents setting forth the terms of the contracts.

allowance   In bidding, money set aside in contracts for items that have not been selected and specified. Bidders are required to include allowances in their proposal and contract amounts. For example, an electrical allowance sets aside an amount of money to be spent on electrical fixtures.

alterations   Partial construction work done within an existing structure without new building addition.

alternates   Amounts stated in bids to be added or deducted from the base bid amount proposed for alternative materials and/or construction methods. The owner is to decide whether an alternate should be incorporated into contract sum at the time of contract award.

application for payment   Contractor's written request for payment for completed work and/or for materials delivered onsite.

approved bidders list   A list of contractors who have met prequalification criteria set by an owner.

approved equal   A contract clause stating that products finally supplied or installed must be equal to originally specified and approved by the architect/engineer.

architect/engineer   The professional hired by an owner to provide design services.

as-built drawings   Drawings marked up to reflect changes made during construction process to contract drawings, showing the locations, sizes, and nature of the building. They are permanent records for future reference.

bid bond   A bond issued by a surety to ensure that, if the bid is accepted, the contractor will sign a contract in accordance with the proposal.

bid documents   Drawings, details, and specifications for a particular project.

bid form   A standard written form furnished to all bidders so they can submit their bids using the same format.

bid opening   The actual process of opening and tabulating bids submitted. It can be open (where bidders are allowed to attend) or closed (where bidders are not allowed to attend).

bid security   A bid bond or certified check, guaranteeing that the bidder will sign a contract, if offered, in accordance with the proposal.

bid shopping   Negotiations between general contractors (buyers) and trade contractors (sellers) to obtain lower prices before/after submitting prime contract proposals to owners.

bid tab   A summary sheet listing all bid prices from contractors or suppliers.

bill of material   A list of items or components used for fabrication, shipping, receiving, and accounting purposes.

bonded roof   A roof carrying a written warranty, usually about weather tightness, including repair or replacement on a prorated cost basis for a certain number of years.

budget estimate   Sometimes called a "ballpark" estimate, it is an estimate based on incomplete information such as schematic drawings.

builder's risk insurance   A form of property insurance covering a project under construction.

building envelope   The building structural framework, also called "Building Shell," mainly consists of exterior cladding and roof system.

CO (Certificate of Occupancy)   A certificate issued by the local municipality after all inspections are completed and all fees paid. It is required before anyone can occupy the constructed facility.

certificate of payment   Statement by an architect informing the owner of the amount due to a contractor based on work completed or materials stored.

change order   A written document signed by owner and contractor authorizing a change in the work or an adjustment in the contract sum or the contract time.

cladding   The external covering to the frame or structural walls of a building. It is designed to carry only its own weight and limited loads such as wind and seism. It can be either fully bonded with the structure it encloses or separate from it by an air barrier.

claim   A formal notice sent by a contractor to an owner requesting additional compensation or an extension of time.

closed bid   A process where only invited bidders are allowed to submit proposals.

commencement of work   The date when a written notice to proceed is sent from an owner to a contractor.

commissioning   When a project is near completion, the constructed facility is put into use to see if it functions as designed.

conditions of the contract   General, Supplementary and Special Conditions of a construction contract.

construction budget   The target cost figure covering the construction phase of a project. It does not include cost of land, A/E design fees, or consultant fees.

construction schedule   A graphic, tabular, or narrative representation of project construction phases, showing activities and durations in sequential order.

contingencies   Amounts in a project budget dedicated to specific cost areas where oversight is an inevitable problem.

contract documents   A term used to represent all agreements between owner and contractor, any general, supplementary or special conditions, working drawings and specifications, all addenda, and post-award change orders.

contract overrun   The cost difference between the original contract price and the final completed cost including all adjustments made by approved change orders.

contract sum   The total agreeable amount payable by an owner to a contractor for the performed work under contract documents.

contractor's option   A written clause in contract documents giving a contractor the option of choosing certain specified materials, methods, or systems without changing the contract sum.

cost breakdown   A breakdown furnished by a contractor describing portions of the contract sum allocated for principal divisions of the work.

cost codes   A numbering system given to specific portions of work to control construction costs.

cost plus contract   A form of contract under which a contractor is reimbursed for direct and indirect costs and is paid a fee for services rendered. The fee is usually stated as a stipulated sum or as a percentage of cost.

critical path method (CPM)   A scheduling diagram drawn to show the tasks involved in constructing a project. Critical path refers to the continuous chain of activities from project start to project finish, whose durations cannot be exceeded in order to complete the project on time.

CSI master format   A system of numbers and titles for organizing construction information into a standard order or sequence.

dead load   The weight of the structure itself, including floor, roof, walls, and any permanent loads.

deflect   To bend or deform under weight.

demising walls   The boundaries that separate a space from neighbors' and from the public corridor.

density   The weight of material in a unit volume.

design-build construction   A contractor bids or negotiates to provide design and construction services for the entire construction project.

design pressure   Specified pressure a product is designed to withstand.

direct costs   The costs directly attributed to a work scope, such as labor, material, and subcontracts. It does not include indirect costs such as office overhead.

direct labor costs   Costs from direct labor hours, not including the add-on portions such as overtime, insurances, and payroll taxes.

direct material costs   Costs for buying materials, including purchase price, freight, and taxes.

draw   The amount of progress billings on a contract that is currently available to a contractor.

factory mutual (FM)   An insurance agency that has established strict construction guidelines as relates to fire and environmental hazards.

fast-track   The process for a contractor to start the construction work before plans and specifications are complete.

final acceptance   The owner's acceptance of a completed structure from a contractor.

final inspection   A final site review by an owner before issuing the final payment.

final payment   The last payment from an owner to a contractor for the entire unpaid balance of the contract sum as adjusted by any approved change orders.

fixed fee contract   A contract with fixed amounts for overhead and profit for all work performed.

fixed price contract   A contract with a set price for the work, same as "lump sum."

gantt chart   A project schedule that shows start and finish dates, critical and noncritical activities, slack time, and relationships among activities.

general conditions   A written portion of contract documents stipulating contractor's minimum acceptable performance requirements.

general contractor (or prime contractor)   A properly licensed individual or company that takes full responsibility for project completion, although he may also enter into subcontracts with others for specific portions of the project. A general contractor assembles and submits a proposal for the work on a project and then contracts directly with the owner to construct the project.

Guaranteed Maximum Price (GMP)   A form of contracts where a contractor guarantees a max ceiling price to the owner for construction cost.

hard costs   Costs directly associated with construction.

indirect costs   Costs for items and activities not directly related to constructing a structure but are necessary to complete the project, for example, contractor's overhead expense.

impact resistance   Ability to withstand mechanical or physical abuse under severe service conditions.

indemnification clause   Provision in a contract in which one party agrees to be financially responsible for specified types of damages, claims, or losses.

INR (Impact Noise Rating)   A single figure rating which provides an estimate of the impact sound insulating performance of a floor-ceiling assembly.

instruction to bidders   A document stating the procedures to be followed by bidders.

invitation to bid   An invitation to a selected list of contractors including bid information related to their trades.

jobsite overhead   Necessary onsite construction expenses, such as construction support costs, supervision, bonus labor, field personnel, and jobsite office expenses.

letter of intent (notice of award)   A notice from an owner to a contractor stating that a contract will be awarded to this contractor. It usually functions as a formal notice to proceed on the project.

lien   The right to hold or sell an owner's property for payment of debts to contractors.

lien release   A written document from contractor to owner that releases the lien following its satisfaction.

lien waiver   A written statement from a contractor or material supplier giving up lien rights against an owner's property.

life-cycle cost   The cost of purchasing, installing, owning, operating, and maintaining a construction element over the life of a facility.

liquated damages   An agreed-to amount chargeable against a contractor as reimbursement for damages suffered by the owner because of the contractor's failure to meet obligations (e.g., not meeting the deadline).

live load   Loads produced by use and occupancy of a building and do not include construction or environmental loads. Examples include people and movable equipment.

load-bearing wall   A strong wall that can support loads in addition to its own weight.

long-lead items   Construction material and equipment which take significant time in fabrication and delivery to jobsite from their purchase dates, for example, structural steel or elevators.

lump sum contract   A written contract between the owner and contractor wherein the owner agrees the pay the contractor a fixed sum of money as a total payment for completing a scope of work.

manufacturer's specifications   The written installation and maintenance instructions developed by the manufacturer of a product and have to be followed in order to maintain the product warranty.

maximum occupancy load   The maximum number of people permitted in a room and is measured per foot for each width of exit door. The maximum is 50 per foot of exit.

mockup   Small scale demonstration of a finished construction product.

monolithic   Single piece of material formed without joints or seams.

notice to proceed   A notice from an owner directing a contractor to begin work on a project.

open bid A bid where any qualified bidder or estimator is given access to project information and allowed to submit a proposal.

owner-builder A term used to describe an owner who takes the role of general contractor to build a project.

payment bond A written form of security from a surety company to the owner, on behalf of a contractor, guaranteeing payment to all persons providing labor, materials, equipment, or services in accordance with the contract.

penalty clause A contract provision that provides for a reduction in payment to a contractor as a penalty for failing to meet deadlines or requirements of contract specifications.

performance bond A written form of security from a surety company on behalf of contractor to the owner that guarantees the contractors' proper and timely completion of a project.

performance specifications Written material containing the minimum acceptable quality standards necessary to complete a project.

permeability A measure of water vapor movement through a material.

permit Written authorization from local governments giving permission to construct or renovate a building. Types of permits include zoning/use permit, demolition permit, grading permit, septic permit, building permit, electrical permit, and plumbing permit.

phased construction Constructing a facility in separate phases. Each separate phase is a complete project in itself.

plan submittal Submission of construction plans to the city or county in order to obtain a building permit.

plan view Drawing of a structure with the view from overhead, looking down.

plot plan A bird's-eye view showing how a building sits on its lot, typically showing setbacks (how far the building must sit from the road), easements, rights of way, and drainage.

porosity The density of substance and its capacity to pass liquids.

preliminary drawings The drawings that precede the final approved drawings.

prequalification A screening process of perspective bidders where an owner gathers background information for selection purposes. Considerations include competence, integrity, reliability, responsiveness, bonding capacity, and similar project experience.

product data Detailed information provided by material and equipment suppliers showing that the item provided meets the requirements of contract documents.

progress payment Periodical payments to a contractor, usually based on the amount of work completed or material stored. There may be a temporary "retainage" or "holdback" (e.g., 5% of the total value completed) which are to be refunded at the end of the project.

project directory A written list of names, addresses, telephone and fax numbers for all parties connected with a specific project, which includes owner, architect, attorney, general contractor, civil, structural, mechanical and electrical engineers.

project manual An organized book that contains general conditions, supplementary and special conditions, the form of contract, addenda, change orders, bidding requirements, proposal forms, and the technical specifications.

project representative A qualified individual authorized by the owner to assist in the administration of a construction contract.

proposal A written offer from a bidder to the owner to perform the work per specific prices and terms.

proposal form   A standard written form furnished to all bidders to submit proposals. It requires signatures from the authorized bidding representatives.

punch list   A list of defects prepared by the owner that need to be corrected by the contractor immediately.

qualification statement   A written statement of the contractor's experience and qualifications during the contractor selection process.

red line   Blueprints that reflect changes and are marked with red pencil.

reimbursable expenses   Expenses that are to be reimbursed by the owner.

relative humidity   The amount of water vapor in the atmosphere at a given temperature.

retainage   Also called "holdback," an amount withheld from each payment to the contractor per Owner-Contractor Agreement.

RFI (Request for Information)   A written request from a contractor to the owner or architect for clarification or information about the contract documents.

RFP   Request for Proposal.

schedule of values   The breakdown of a lump sum price into smaller portions of identifiable construction elements, which can be evaluated for progress payment purposes.

separate contract   A contract between an owner and a contractor, other than the general contractor, for constructing a portion of a project.

settlement   Downward movement of the soil or of a structure which it supports.

shop drawings   Drawings that show how specific portions of the work will be fabricated and installed.

soft costs   Cost items in addition to direct construction costs. They generally include architectural and engineering fees; legal, permits, and development fees; financing fees; leasing and real estate commissions; advertising and promotion and so on.

special conditions   Amendments to the general conditions that change standard requirements to unique requirements for a specific project.

specifications   Detailed statements describing materials, quality, and workmanship to be used in a specific project. Most specifications use the Construction Specification Institute (CSI) format.

standard details   A drawing or illustration sufficiently complete and detailed for use on other projects with minimum or no changes.

startup   The period prior to owner occupancy when mechanical, electrical, and other systems are activated and the owner's operating/maintenance staff is instructed in their use.

subcontractor bond   A written document from a subcontractor given to the general contractor, guaranteeing performance of the contract and payment of all labor, materials, and service bills associated with the subcontract agreement.

substantial completion   The stage when the construction is sufficiently complete and in accordance with the contract documents, so the owner may occupy the facility for the intended use.

substitution   A proposed replacement offered in lieu of and as being equivalent to a specified material or process.

substructure   The supporting part of a structure, that is, the foundation.

superstructure   The part of a building above the foundation.

supplemental conditions   A written section of the contract documents supplementing or modifying the standard clauses of general conditions.

square   A 10 ft × 10 ft area, or 100 square feet, usually applied to roofing and siding material.

STC (Sound Transmission Class)   The measure of sound stopping of ordinary noise. It is used for interior walls, ceilings, and floors.

stop order   A formal, written notification to a contractor to discontinue work on a project for reasons such as safety violations, defective materials or workmanship, or cancellation of the contract.

structural concrete   Reinforced concrete with a compressive strength of at least 2,500 psi.

subcontractor   A contractor hired by the general contractor to engage in a particular trade such as steel erection.

takeoff   Figuring materials necessary to complete a job.

TI (Tenant Improvement)   The interior improvement to a building after its shell is complete. It usually includes finish floor coverings, ceilings, partitions, doors with frames and hardware, fire protection, HVAC ductwork and controls, electrical lighting, and ancillary systems. The cost of TI is generally paid by the tenant.

trade contractor   A contractor that specialized in specific elements of the overall project.

transmittal   A written document used to identify what is being sent to a receiving party. The transmittal is usually the cover sheet for the package sent and includes the contact information of the sender and receiver.

travel time   Wages paid to workers under certain union contracts and under certain job conditions for the time spent in traveling between their residence and the job.

UL (Underwriters' Laboratories)   An independent testing agency that checks products and materials for safety and quality.

unit price contract   A written contract where an owner agrees to pay a contractor a specified amount of money for each unit of work completed. The designated unit price is all-inclusive: labor, materials, equipment, overhead and profit.

value engineering   A technical review process aimed to save construction costs by changing design options.

verbal quotation   A written document used by the contractor to record a subcontract or material cost proposal over the telephone, prior to receiving the written proposals via mail, email, or fax later.

void   Cavity in cellular materials such as concrete.

warranty   There are two types of warranties. One is provided by the manufacturer of a product such as roofing material or an appliance. The other is a warranty for the labor. For example, a roofing contract may include a 20 year material warranty and a 5 year labor warranty.

water table   The vertical distance from the earth surface to underground water.

zoning   A governmental process that limits the use of a property for a specific purpose, for example, single-family use, high-rise residential use, or industrial use.

BASICS OF WINDOWS-BASED PROGRAMS

Parts of a Typical Window

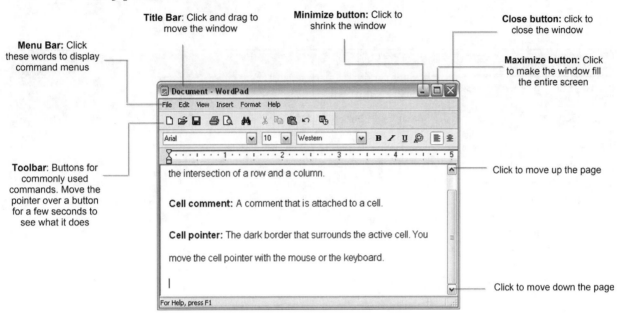

Title Bar: Click and drag to move the window

Minimize button: Click to shrink the window

Close button: click to close the window

Menu Bar: Click these words to display command menus

Maximize button: Click to make the window fill the entire screen

Toolbar: Buttons for commonly used commands. Move the pointer over a button for a few seconds to see what it does

Click to move up the page

Click to move down the page

Document - WordPad

File   Edit   View   Insert   Format   Help

Arial   10   Western   **B** *I* U

the intersection of a row and a column.

Cell comment: A comment that is attached to a cell.

Cell pointer: The dark border that surrounds the active cell. You

move the cell pointer with the mouse or the keyboard.

For Help, press F1

File Management

Click a folder in the left pane ......to view contents in the right pane

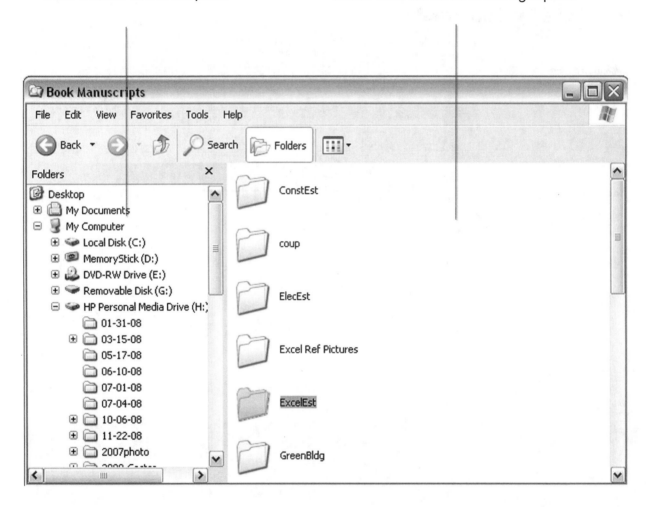

File Operations	What to Do
Open a file or folder	Double-click on it
Create a folder	Go to menu *File → New → Folder*
Rename a file	Select it, go to menu *File → Rename*
Make a copy of a file	Select it, go to menu *Edit → Copy to Folder*
Copy a file to a removable disk	Select it and right-click. Pick *Send To* from the shortcut menu. Pick your removable disk.
Select multiple files	Hold down CTRL key while clicking files one by one.
View a file's properties	Select it and right-click. Pick *Properties*.
Find a file	Click *Start* menu button on the lower left of your screen and select *Search*.

EXCEL TOOLBAR BUTTONS

Standard Toolbar

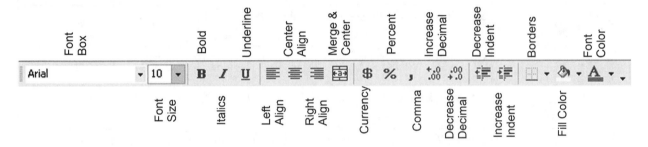

Formatting Toolbar

Standard Toolbar Buttons

Button	Description
	New: Open a new, blank workbook. Any current workbooks will remain open in the background.
	Open: Open an existing workbook on your computer or network.
	Save: Save the current workbook.
	Print: Print a copy of the current worksheet or print area using the default settings.
	Print Preview: Activate Print Preview mode, which lets you see how your worksheet or print area will look on the printed page.
	Spelling: Check the spelling of all worksheet cells.
	Cut: Remove the selected object and place it on the Clipboard for later use.
	Copy: Place a copy of a selected object on the Clipboard for later use.
	Paste: Place a copy of the object currently stored on the Clipboard into the workbook.
	Format Painter: Copy the formatting of the selected object and apply the same formatting to other objects.
	Undo: Reverse an action that you have just performed. Use the down-arrow to open a list of actions that can be undone simultaneously.
	Redo: Restore an action that you have just undone. Use the down-arrow to open a list of actions that can be restored simultaneously.
	Insert Hyperlink: Insert or edit a hyperlink that jumps to another document, another application's file, or another Web page.
Σ	**AutoSum:** Insert the SUM function that totals a proposed cell range.
f_x	**Paste Function:** Open the Paste Function dialog box to begin creating a function.
	Sort Ascending: Sort the selected rows in ascending order.

Button	Description
Z A↓	**Sort Descending:** Sort the selected rows in descending order.
(chart wizard icon)	**Chart Wizard:** Create or modify a chart.
(drawing icon)	**Drawing:** Show or hide the Drawing toolbar.
100% ▼	**Zoom:** Change how large or small the worksheet appears on the screen. Use the down-arrow to open a list of preset options.
[?]	**Office Assistant:** Activate the Office Assistant to get help or tips as you work with Excel.

Formatting Toolbar Buttons

Button	Description
Arial ▾	**Font List Box:** Apply an installed font to the cell contents. Use the down-arrow to open a list of fonts.
10 ▾	**Font Size List Box:** Apply a font size to the cell contents. Use the down-arrow to open a list of sizes or type a size.
B	**Bold:** Apply bold formatting to the cell contents.
I	**Italic:** Apply italic formatting to the cell contents.
U	**Underline:** Apply underline formatting to the cell contents.
≣	**Align Left:** Align the cell contents to the left edge.
≣	**Center:** Center the cell contents in the selected cell.
≣	**Align Right:** Align the cell contents to the right edge.
⊞	**Merge and Center:** Merge the selected cells and center the contents in the new cell.
$	**Currency Style:** Apply currency format to the cell contents.
%	**Percent Style:** Apply a percent format to the cell contents.
,	**Comma Style:** Apply comma format to the cell contents.
$+.0 \atop .00$	**Increase Decimal:** Add one decimal place to the cell contents.
$.00 \atop +.0$	**Decrease Decimal:** Remove one decimal place.
⇤	**Decrease Indent:** Decrease the indent level of the cell contents by one character width.
⇥	**Increase Indent:** Indent the cell contents by one character width.
⊡	**Border:** Apply the current border to the selected cells. Use the down-arrow to select among different border styles.
⬙	**Fill Color:** Apply the current color to the *selected cells*. Use the down-arrow to select among different colors.
A	**Font Color:** Apply the current color to the *cell contents*. Use the down-arrow to select among different colors.

EXCEL KEYBOARD SHORTCUTS
General

Action	Keystrokes
Create a new workbook	CTRL + N
Open a workbook	CTRL + O
Save a workbook	CTRL + S
Save a workbook with a different name	F12
Search for a workbook	ALT + F, then press H
Print review	ALT + F, then press V
Set print area	ALT + F, press T, then press S
Page setup	ALT + F, then press U
Print	CTRL + P
Protect a workbook/worksheet	ALT + T, then press P
E-mail a workbook	ALT + F, press D, then press A
Close a workbook	CTRL + W
Quit Excel	ALT + F4
Help	F1
Display/hide task pane	ALT + V, then press K
Import data into workbook	ALT + D, then press D twice
View workbook properties	ALT + F, then press I
Compare two workbooks	ALT + W, then press B
Switch between applications	ALT + TAB

Navigation

Action	Keystrokes
Between cells	Four arrow keys
Up one screen	Page Up
Down one screen	Page Down
Top of worksheet (cell A1)	CTRL + HOME
End of worksheet	CTRL + END
Beginning of row	HOME
End of row	CTRL + → (right arrow)
Next worksheet	CTRL + Page Down
Previous worksheet	CTRL + Page Up
Go to cell or cell ranges	F5

Change View

Action	Keystrokes
Full screen view	ALT + V, then press U
Zoom	ALT + V, then press Z
Page break view	ALT + V, then press P
Normal view	ALT + V, then press N
Freeze worksheet titles	ALT + W, then press F
Split screen	ALT + W, then press S
Compare two workbooks	ALT + W, then press B
Hide column	CTRL + 0 (number zero key)
Display hidden column	CTRL + SHIFT +)
Hide row	CTRL + 9 (number nine key)
Display hidden row	CTRL + SHIFT + (
Hide worksheet	ALT + O, press H, then press H
Display hidden worksheet	ALT + O, press H, then press U
Autofilter	ALT + D, then press F twice
Sort the data	ALT + D, then press S
Group and outline	ALT + D, then press G
Header and footer	ALT + V, then press H

Selection

Action	Keystrokes
All cells left of current cell	SHIFT+ ←
All cells right of current cell	SHIFT+ →
First cell	CTRL + HOME
Last cell	CTRL + END
Content in the cell	Double-click, then select
Adjacent cells	Click first cell, press SHIFT, then click last cell
Nonadjacent cells	Hold down CTRL key while clicking cells one by one
Entire column	CTRL + SPACEBAR
Entire row	SHIFT + SPACEBAR
Entire worksheet	CTRL + A

Edit Data

Action	Keystrokes
Insert new worksheet	SHIFT + F11
Edit active cell	F2
Insert row	ALT+ I, then press R
Insert column	ALT+ I, then press C
Cut	CTRL + X
Copy	CTRL + C
Paste	CTRL + V
Undo	CTRL + Z
Redo	CTRL + Y
Find	CTRL + F
Replace	CTRL + H
Paste special	ALT+ E, then press S
Clear cell content	DELETE
Spell check	F7
Insert comments	SHIFT + F2
Insert picture	ALT + I, then press P
Insert hyperlink	CTRL + K
Shortcut menu	SHIFT + F10
Make a chart	F11

Formulas

Action	Keystrokes
Apply AutoSum	Press ALT key and = (equal sign) key at once
Edit a formula	F2
Insert function	SHIFT + F3
Insert current date	CTRL + ; (semicolon key)
Insert current time	CTRL + : (colon key)
Make absolute reference	F4
Define a name	CTRL + F3
Error checking	ALT + T, then press K
Help	F1
Record macro	ALT + T, then press M, then press R
Execute macro	ALT + F8

Format Data

Action	Keystrokes
Format cells dialog box	CTRL + 1 (number one key)
Format as general number	SHIFT + CTRL + ~
Format as number with two decimal places and thousand separator	SHIFT + CTRL + !
Format as currency	SHIFT + CTRL + $
Format as percent	SHIFT + CTRL + %
Format as date	SHIFT + CTRL + #
Format as time	SHIFT + CTRL + @
Apply cell outline border	SHIFT + CTRL + &
Remove cell outline border	SHIFT + CTRL + _
Make text bold	CTRL + B
Make text italics	CTRL + I
Make text underline	CTRL + U
Make text strikethrough	CTRL + 5
Change font type	CTRL + SHIFT + F, then ALT + ↓
Change font size	CTRL + SHIFT + P, then ALT + ↓
Color cell background	CTRL + 1, then press P
Cell text position, wrapping or merge	CTRL + 1, then press A
Set column width	ALT + O, press C, then press W
Set row height	ALT + O, press R, then press E
AutoFormat	ALT + O, then press A
Apply format style	ALT + O, then press S
Conditional formatting	ALT + O, then press D
Set picture as sheet background	ALT + O, press H, then press B

EXCEL FORMULAS AND FUNCTIONS

Formula Basics

Formulas are used to calculate new values from existing values in a worksheet.

The following example adds 25 to the value in cell B4, and then divides the result by the sum of the values in cells D5, E5, and F5.

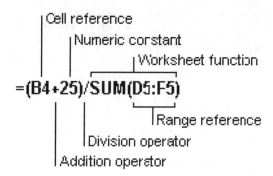

When the input cell values change, Excel will update the calculation result automatically.

Enter Formulas

Step 1: Select the cell and type an equal sign (=).

Step 2: Enter raw numbers (like 123), operators (like +, −), cell references (like A1), or functions (like Sum [B2:B9]). Most math symbols are on the small 10-key pad (+ for addition, * for multiplication, / for division, − for subtraction).

Step 3: Press ENTER key.

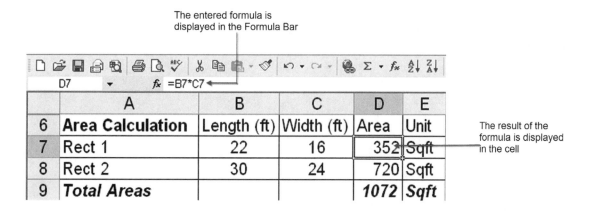

Cell References

Cell references are those letters and numbers that refer to a cell's location (e.g., A4 or C5:C7). Formulas can contain cell references as well as raw numbers. For example, if you type **=B7+C7,** Excel will add the values in cell B7 and cell C7. Cell reference is usually a good method to use for formulas because if you change the values in cells referenced by a formula (in this case B7 and C7), the calculated result updates automatically.

There are two types of operators for cell reference. Union operator, the comma (,), only refers to two individual cells. Range operator, the colon (:), refers to all cells between the two cells specified.

Cell References	Refer to Values in:
A10	the cell in column A and row 10
A10,A20	cell A10 and cell A20
A10:A20	the range of cells in column A and rows 10 through 20
B15:E15	the range of cells in row 15 and columns B through E
A10:E20	the range of cells in columns A through E, and rows 10 through 20

Enter Cell References in Formulas

To enter a cell reference in a formula, type it in directly, or point and click on the cell and then Excel will fill in the cell reference. Input a formula by pointing is recommended, because it is more visual and less error-prone.

1. **By Typing** (works best for easy formulas):
    a. Select the cell where you want the formula results to appear.
    b. Press the equal sign = to designate the entry as a formula.
    c. Type the appropriate cell references and mathematical operators.
    d. Press ENTER key.
    Example: Directly typing the formula **=C7+D7** from left to right.

2. **By Pointing** (Useful for complicated formulas):
    a. Select the cell where you want the formula result to appear.
    b. Press the equal sign (=) to designate the entry as a formula.
    c. Click (or drag if cells are adjacent) your mouse on each cell to place its reference in the formula bar.
    d. Type math operators where needed.
    e. Press ENTER key.
    Example: To put the same formula **=C7+D7,** type in an equal sign (=), click cell *C*7, then type the plus sign, and finally click cell *D*7.

Formula Operators

Operators are math symbols used to perform calculations in formulas, such as a plus sign (+). When cells contain numerical data, you can add, subtract, multiply, and divide the cell contents.

Operations	Operator	Example	Meaning
Addition	+ (plus sign)	**=10+13**	Adds 10 to 13
		=B1+B2	Adds the values of cells B1 and B2
Subtraction	− (minus sign)	**=C9−B2**	Subtract the value of cell B2 from that of cell C9
Multiplication	* (asterisk)	**=C8*B9**	Multiply the values of cell C8 and B9
Division	/ (forward slash)	**=15/3**	Divide 15 by 3
Percent	% (percent sign)	**=20%*B2**	Take 20% from the value of cell B2
Exponentiation	^ (caret)	**=A5^3**	Raise cell A5 value to power 3 (e.g., 2^3=8)
Joining Text	& (ampersand without dot)	**= "Adam" & "Ding"**	Join the two texts into one (i.e., "AdamDing")

Order of Operator Precedence

Formula calculations are carried out from left to right if operators have the same level of precedence. Multiplication and division are on the same level. Addition and subtraction are on the next lower level.

When a formula has several operators, **PEMDAS** represents the order is which the calculations are performed.

- Parentheses: All calculations inside parentheses are performed first.
- Exponents: All values with exponents are calculated.
- Multiplication and Division: Any multiplication or division is performed.
- Addition and Subtraction: Any addition or subtraction is performed.

An easy way to remember this order is: **Please** (*parentheses*) **excuse** (*exponents*) **my** (*multiplication*) **dear** (*division*) **aunt** (*addition*) **Sally** (*subtraction*).

In the formula **=(B4+25)/SUM(D5:F5),** the parentheses around the first part of the formula force Excel to calculate B4+25 first and then divide the result by the sum of the values in cells D5, E5, and F5.

Function Basics

Functions are predefined worksheet formulas. A function consists of a name followed by specific values, called arguments. For example, the following SUM worksheet function adds values of cells A10, B5 through B10, numbers 50 and 37.

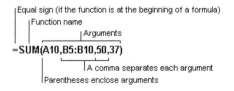

A function has three parts:

1. Equal sign
2. Function name
3. Argument, enclosed in parentheses and separated by commas

SUM Worksheet Function

SUM worksheet function is the most frequently used function. It can be activated with the following steps:

1. Click a cell below the column of values or to the right of the row of values.
2. Click the AutoSum button **Σ** on standard toolbar. If the range that Excel suggests for the sum is wrong, drag your mouse to correct it.
3. Press ENTER key

Example: Enter formula: **=SUM (B2:B7)**

To add nonadjacent values in a column or row:

1. Type an equal sign **=**
2. Type **SUM**
3. Type an opening parenthesis **(**
4. Type or select the cell references you want to add. Use comma to separate individual arguments that tell the function what to calculate.
5. Enter a closing parenthesis **)**
6. Press ENTER Key

Example: Enter formula: **=SUM(B2,B5,B7)**

Common Worksheet Functions

Common worksheet functions include the following:

Function	Meaning
SUM	Add a group of numbers
AVERAGE	Average a group of numbers
MAX	Find the largest value
MIN	Find the smallest value
IF	Function value based on if a test condition is true or false
VLOOKUP	Find a specified value in a table
ROUNDUP	Round up a number with specified decimals
SUMIF	Add up the cells specified by a given criteria
SQRT	Take the square root of a number
COUNT	Count the number of items
UPPER	Convert all text to uppercase
LOWER	Convert all text to lowercase
TRIM	Removes extra spaces from a text

There are hundreds of other functions. To explore them, click on the Function Wizard button $f_{\!x}$, and the following screen will appear.

Enter a description of the function you want to use

Click to search for a function

Select one of Excel's 10 function categories

List of functions for each category

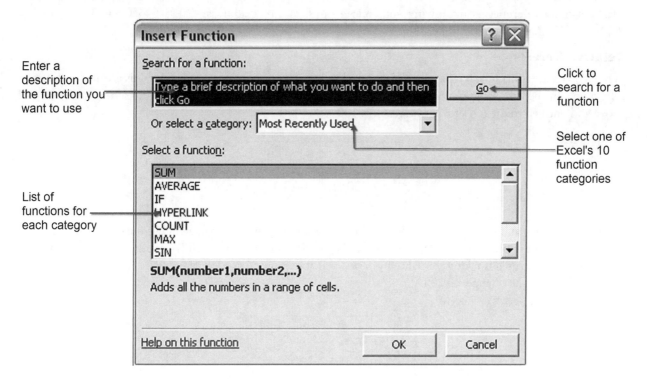

Copy a Formula or Function

Copy a formula into adjacent cells using the AutoFill handle using these steps.

1. Select the cell that contains the formula you want to copy.
2. Position the mouse pointer over the lower-right corner of the cell until the black cross or AutoFill handle (+) appears.
3. Drag the AutoFill handle 🖱️ down the column, or across rows.

Copy a formula into nonadjacent cells using these steps.

1. Select the cell that contains the formula you want to copy from.
2. Press CTRL + C on keyboard to copy.
3. Select the cell or cells that you want to paste the formula to.
4. Press CTRL + V on keyboard to paste.

By default, copying and pasting formulas is by "relative reference." That means cell references in these formulas will be changed during the process. If you do not want them to change, use "absolute reference" (see next section for details).

Relative Reference Versus Absolute Reference

Relative reference is like giving someone directions from a present location: "Go north two blocks and turn left." Absolute reference is like giving someone a physical street address: "#123-914 Craig Rd, Kelowna, BC V1X 7Z7, Canada."

Relative Reference

When you create a formula, it tells Excel how to find cells starting from the current cell. This is called as "relative reference". In the following example, cell B6 contains the formula 5A5. This means it takes the value one cell above and one cell to the left of B6. The result is 100.

	A	B
5	100	
6	200	=A5

When you copy and paste a formula based on relative reference, Excel automatically adjusts cell references. In the above same example, when you copy the formula (**=A5**) from cell B6 to cell B7, it automatically changes to **=A6.** The new formula is taking the value from one cell above and to the left of cell B7 (i.e., taking the value of cell A6). So the result changes to 200.

	A	B
5	100	
6	200	=A5
7		=A6

Absolute Reference

When you create a formula, you can tell Excel to find cells based on an exact location that never moves. This is called as "absolute reference". In the following example, cell B6 contains the formula **=$A$5.** You need to put dollar signs before cell references in order to make it work. This means it always takes the value of current cell A5 (i.e., 100).

	A	B
5	100	
6	200	=$A$5

When you copy and paste a formula based on absolute reference, Excel will NOT adjust cell references. In the above same example, when you copy the formula (**=$A$5**) from cell B6 to cell B7, it remains as **=$A$5.** So the result is still 100.

	A	B
5	100	
6	200	=$A$5
7		=$A$5

Mixed Reference

Mixed references are a blend of absolute and relative references, which looks like A$1 or $A1. The purpose is to allow only column (or row) to vary when copying formulas, while freezing row (or column). This category is left to interested readers themselves.

Auto Calculate: See the Result Without Entering Formula

What is the total of these numbers? At least you know where to find the AutoSum button Σ on the toolbar. If cells are not next to each other, you need to manually typing a formula to add them up. But there is also a way that you can see the addition result without messing with formulas.

	A	B	C	D
1	Items	Strength	QTY	Unit
2	Footing	3000 PSI	33.4	CY
3	Pads	3000 PSI	11.2	CY
4	Beams	4000 PSI	2.5	CY
5	Columns	4000 PSI	5.5	CY
6	Stairs	3500 PSI	2.1	CY
7	Slabs	3000 PSI	50.3	CY

For example, if you want to know how many cubic yards of foundation and slabs are in the above worksheet, select cells C2, C3, and C7 (press CTRL key when picking). Then look at the lower right corner of your screen, on the Status toolbar (the very bottom one) for the total of three cells: "SUM = 94.9". It looks like this: **Sum=94.9**

This feature is called "AutoCalculate." The example shows the sum result without entering a formula for adding. If you right click on "SUM" word on the Status toolbar, it actually gives you more options to find out the average, maximum or minimum value of the selected cells. Interested readers can explore further.

Formula Errors

Excel needs precise instructions. So you should enter formulas exactly as required. Missing a comma or parenthesis, inserting an extra space, or misspelling a function name could produce errors. For example, misspelling a worksheet function name will generate the *#NAME?* error, instead of the correct result you wanted.

Common formula errors include the following:

Error Messages	Meaning
#####	Cell contents too wide for display (This is not really an error)
#DIV/0!	Divided by zero or an empty cell
#REF!	Invalid cell reference (cells deleted or relocated)
#VALUE	Wrong type of data (e.g., adding texts to numbers)
#NUM	Invalid number (the result too large or too small)
#Name?	Misspelled function name or cell reference
#Null!	Two cell areas do not intersect
#N/A	Result not available
Circular Reference	A cell reference in a formula refers to the formula's result

Ways to reduce formula errors:

- Remember the equal sign (=) at the start of the formula.
- Do not put an extra space before the equal sign.
- Have your data in the right format (e.g., all math calculations should be performed on numbers instead of texts).
- Select the correct cell range.
- Spell function names correctly.
- Place parentheses around function arguments.
- Separate function arguments inside parentheses with commas.

SPREADSHEET TERMS

absolute reference   In a formula, a reference to a cell that does not change if the formula is copied to a different cell. An absolute reference uses two dollar signs, such as $A$15 for cell A15.

active cell   The selected cell where data is entered when you begin typing. Only one cell is active at a time. The active cell is bounded by a heavy border and its contents appear in the formula bar.

argument   In a worksheet function, information (enclosed in parentheses) that provides details as to what you want the function to do.

cell   The fundamental data storage unit in a worksheet, defined by the intersection of a row and a column.

cell comment   A comment that is attached to a cell.

cell pointer   The dark border that surrounds the active cell. You move the cell pointer with the mouse or the keyboard.

cell reference   Identifies a cell by giving its column letter and row number. For example, C5 refers to the cell at the intersection of column C and row 5. If you are referring to a cell on a different sheet, you need to precede it with the sheet name and an exclamation point. These can be relative references (most common), absolute references, or mixed references.

chart   A graphic representation of values in a worksheet. A chart can be embedded on a worksheet or stored on a separate chart sheet in a workbook.

circular reference   In a formula, a reference to the cell that contains the formula (either directly or indirectly). If cell A10 contains **=SUM(A1:A10),** a circular reference exists because the formula refers to its own cell.

column   Part of a worksheet that consists of 65,536 cells arranged vertically. Each worksheet has 256 columns.

conditional formatting   Formatting (such as color or bold text) that is applied to a cell depending on the cell's contents.

data validation   The process of ensuring that data of the correct type is entered into a cell. For example, if the entry is outside of a specified range of values, you can display an error message to the user.

fill handle   The small square object that appears at the lower-right corner of the active cell or a selected range of cells.

formula   An entry in a cell that returns a calculated result.

formula bar   The area of Excel, just below the toolbars, that displays the contents of the active cell. You can edit the cell in the formula bar.

freeze pane   The process of keeping certain top rows and/or left columns always displayed, no matter where the cell pointer is.

function   A special keyword used in a formula to perform a calculation. Use the Function Wizard to enter a function in a formula.

link formula   A formula that uses a reference to a cell that is contained in a different workbook.

number format   The manner in which a value is displayed. For example, you can format a number to appear with a percent sign and a specific number of decimal places. The number format changes only the appearance of the number (not the number itself).

named range   A range that you have assigned a name to. Using named ranges in formulas makes your formulas more readable.

noncontiguous range   A range of cells that are not next to each other. You select a noncontiguous range by pressing CTRL key while you select cells.

merged cells   Cells that have been combined into one larger cell that holds a single value.

operator   In a formula, a character represents the type of operation to be performed. Operators include + (plus sign), / (division sign), and others.

mixed reference   In a formula, a reference to a cell that is partially absolute and partially relative. A mixed reference uses one dollar sign, such as A$15 for cell A15. In this case, the column part of the reference is relative; the row part of the reference is absolute.

pane   One part of a worksheet window that has been split into either two or four parts.

pointing   The process of selecting a range using either the keyboard or the mouse. When you need to enter a cell or range reference into a dialog box, you can either enter it directly or point to it in the worksheet.

range   A collection of two or more cells. Specify a range by separating the upper-left cell and the lower-right cell with a colon.

recalculate   To update a worksheet's formulas using the most current values.

relative reference   In a formula, a reference to a cell that changes (in a relative manner) if the formula is copied to a different cell. A relative reference does not use dollar signs (as opposed to an absolute reference or a mixed reference).

row   Part of a worksheet that consists of 256 cells arranged horizontally. Each worksheet has 65,536 rows.

sheet   One unit of a workbook, which can be a worksheet or a chart sheet. Activate a sheet by clicking its sheet tab.

spreadsheet   A generic term for a product such as Excel that is used to track and calculate data. Or, this term is often used to refer to a worksheet or a workbook.

status bar   The line at the bottom of the Excel window that shows the status of several things and also displays some messages.

value   A number entered into a cell. It can be the number you typed or the calculation result of a formula.

wizard   A series of dialog boxes that assist you in performing an operation such as creating a chart, importing text, or creating certain types of formulas.

workbook   The name for a file that Excel uses. A workbook consists of one or more sheets.

worksheet   A sheet in a workbook that contains cells.

Adam Ding is a professional estimator with extensive experience in a variety of residential, commercial, institutional, industrial, and infrastructure projects. He holds a Master's degree in Building Construction from Auburn University.

Adam has had a successful career in estimating projects of different sizes, ranging from large-scale cost planning to detailed trade takeoff. Having bid countless construction jobs, Adam developed a unique and easy-to-understand estimating approach. He taught classes at Auburn University and also owns the copyrights for a number of spreadsheet estimating programs.

Currently, as a certified Professional Quantity Surveyor (PQS), Adam continues to provide cost estimating services in North America and Asia Pacific regions. He is a regular technical contributor to *Construction Economist* journal, introducing the latest estimating technologies and techniques. Interested readers can also check out Adam's other popular estimating books:

- *DeWalt Construction Estimating Professional Reference*
- *DeWalt Plumbing Estimating Professional Reference*
- *DeWalt HVAC Estimating Professional Reference*
- *DeWalt Electrical Estimating Professional Reference* (2nd edition)